Advances in Geographic Information Science

The series aims to: present current and emerging innovations in GIScience; describe new and robust GIScience methods for use in transdisciplinary problem solving and decision making contexts, illustrate GIScience case studies for use in teaching and training situations, analyze GIScience tools at the forefront of scientific research, and examine the future of GIScience in an expanding knowledge-based economy. The scope of the series is broad and will encompass work from all subject disciplines that develop and use an explicitly spatial perspective for analysis and problem-solving.

Eric Vaz

Regional and Urban Change and Geographical Information Systems and Science

An Analysis of Ontario, Canada

Springer

Eric Vaz
Department of Geography and Environmental Studies
Toronto Metropolitan University
Toronto, ON, Canada

ISSN 1867-2434 ISSN 1867-2442 (electronic)
Advances in Geographic Information Science
ISBN 978-3-031-24730-9 ISBN 978-3-031-24731-6 (eBook)
https://doi.org/10.1007/978-3-031-24731-6

This Springer imprint is published by the registered company Springer Nature Switzerland AG
The registered company address is: Gewerbestrasse 11, 6330 Cham, Switzerland

Contents

1 Land-Use and Land Cover Change: Advancing with Geographic Information Science 1
1.1 Land Use in Ontario 1
1.2 Ontario's Greenbelt 3
1.3 Ontario's Land-Use Organizations 7
1.4 The Importance of Regional Models for Land Use 8
1.5 Governmental Initiatives of Land-Use Collection 9
1.6 Urban Land-Use Planning in Ontario 11
1.7 Volunteered Geographic Information as Land-Use Sources 13
1.8 Conclusions and Future Work 14
References 14

2 Urban Sprawl into the Niagara Region: Urban Encroachment on Agriculture 17
2.1 Introduction 17
2.2 Study Area 18
2.3 Protection Groups and Policies 20
2.3.1 Niagara Escarpment Biosphere Reserve 20
2.3.2 Niagara Escarpment Planning and Development Act (NEPDA) 21
2.3.3 Niagara Escarpment Commission (NEC) 21
2.3.4 Niagara Escarpment Plan (NEP) 22
2.3.5 Greenbelt Act/Plan 23
2.4 Defining Landscape 23
2.5 Data 25
2.6 Methods 26
2.7 Landscape Metrics 29
2.8 Results 30
2.9 Conclusions 31
References 32

3 COVID-19 in Toronto: Investigating the Spatial Impact of Retailers in the Food Retail and Food Service Sector 35
3.1 Introduction 35
3.2 Literature Review.. 37
3.2.1 Factors Influencing the COVID-19 Pandemic........... 37
3.2.2 COVID-19 Impact on the Economy and Retail.......... 38
3.2.3 Changes in Consumer Shopping Behavior and Consumption 39
3.2.4 Applications and Methods Used 40
3.3 Study Area.. 42
3.4 Data 42
3.4.1 COVID-19 Data Gathering and Pre-processing.......... 42
3.4.2 Retail Data Gathering 45
3.4.3 Socio-economic Data Gathering and Pre-processing 45
3.5 Methodology 46
3.5.1 Kernel Density.. 47
3.5.2 Assessing Spatial Autocorrelation: Using Global Moran's I.. 48
3.5.3 Local Moran's I Statistic 49
3.5.4 Geographically Weighted Regression 49
3.6 Analysis Results.. 50
3.6.1 Spatial Distribution of COVID-19.................... 50
3.6.2 Spatial Distribution of Retailers..................... 52
3.6.3 Cross-Comparative Analysis of COVID-19 and Retailer Distribution 53
3.6.4 Geographically Weighted Regression 57
3.7 Discussion 58
3.7.1 Key Findings 59
3.7.2 Contributions 60
3.7.3 Limitations.. 61
3.7.4 Future Research and Next Steps 61
3.8 Conclusion.. 62
References.. 63

4 Spatial Analysis of Characteristics and Influencing Factors of Killed or Seriously Injured Persons from Motor Vehicle Collisions Within the City of Toronto 65
4.1 Introduction 65
4.1.1 Study Area.. 66
4.2 Literature Review.. 66
4.2.1 Socio-economic Factors........................... 66
4.2.2 Spatial Autocorrelation Techniques.................. 67
4.3 Methodology 68
4.3.1 Data 70
4.3.2 Aggregation Techniques 73

4.3.3 Spatial Autocorrelation: Getis-Ord Gi* Statistic 75
4.3.4 Model Evaluation and Stepwise Regression 76
4.4 Results and Discussion 77
4.4.1 Global Autocorrelation and Temporal Characteristics of Injury Incidents 78
4.4.2 Global Spatial Autocorrelation Trends by Region 79
4.4.3 Spatial Clusters and Quantitative Characteristics of KSI by Region 83
4.4.4 Backward Stepwise Regression 89
4.5 Conclusions 93
Appendix: Linear Feature Definitions (City of Toronto 2020) 94
References 95

5 Socio-spatial Variation in Informal Caregiving to Ontario Seniors 97
5.1 Introduction 97
5.2 Literature Review 98
5.2.1 Definition of Informal Caregiving 98
5.2.2 Importance of Informal Senior Caregiving 98
5.2.3 Policy Context 100
5.2.4 Factors Influencing the Provision of Informal Care. 105
5.2.5 Need for a Spatial Understanding of Informal Caregiving 108
5.2.6 GIS Applications in the Continuum of Care for the Elderly 110
5.3 Study Area 111
5.4 Methods and Data 112
5.4.1 Methods 112
5.4.2 Data 116
5.5 Results and Discussion 118
5.5.1 Results 118
5.5.2 Discussion 125
5.6 Conclusion 130
5.6.1 Findings 130
5.6.2 Limitations 131
5.6.3 Future Research 132
Appendices 132
Appendix A: Census and Canadian Community Health Survey Questions 132
Appendix B: Language Groups Used to Create INDIV and FAMIL Variables 134
References 135

6 A Spatial Statistical Understanding of Inadequate Dwellings in the City of Toronto ... 139
6.1 Introduction ... 139
6.1.1 Context ... 139
6.1.2 Research Questions and Objectives ... 140
6.2 Literature Review ... 141
6.2.1 History and Background ... 141
6.2.2 Home Characteristics and Dwelling Condition Impact ... 142
6.2.3 Resident Characteristics and Dwelling Condition Impact ... 144
6.2.4 Dwelling Conditions and Interventions ... 145
6.3 Data and Methodology ... 146
6.3.1 Selecting and Transforming Census Data from Statistics Canada ... 146
6.3.2 Methodology ... 148
6.4 Results ... 152
6.4.1 Hierarchal Cluster Results ... 152
6.4.2 Regression Results ... 154
6.4.3 Bivariate Results ... 157
6.4.4 Spatial Autocorrelation and Spatial Findings ... 158
6.5 Discussion ... 168
6.5.1 Regression Model Revelations ... 168
6.5.2 Bivariate Relationships and Comparisons ... 168
6.5.3 Spatial Patterns and Insights ... 169
6.5.4 Limitations ... 171
6.6 Conclusion and Progress ... 172
Appendix (Tables 6.10, 6.11 and 6.12) ... 173
References ... 176

7 Ontario Wetland Policy Analysis ... 181
7.1 Introduction ... 181
7.2 Context ... 182
7.3 Current Approaches to Wetland Management in Ontario ... 183
7.3.1 Provincial Wetland Conservation Acts and Policy Instruments ... 183
7.3.2 Ontario Conservation Authorities: The Conservation Authority Act and the Drainage Act ... 184
7.3.3 Wetland Conservation Major Actions ... 184
7.4 Critique of Current Approaches ... 185
7.4.1 Critique of Provincial Management ... 185
7.4.2 Critique of the Drainage Act: Limitations Faced by Conservation Authorities ... 187
7.4.3 Critique of Major Actions ... 188
7.5 Alternative Approaches ... 189
7.6 Conclusion ... 191
References ... 192

8 The Spatial and Temporal Distribution of Airbnb and Its Potential Impact on the Rental Market: A Case Study of City of Toronto, Ontario, Canada . . . 195
8.1 Introduction . . . 195
8.2 Background . . . 197
8.2.1 Digital Platform Economy and the Sharing Economy . . . 198
8.2.2 Housing and Rental Market in Canada . . . 198
8.2.3 Policy and Regulation . . . 201
8.3 Methodology . . . 202
8.3.1 Data . . . 202
8.3.2 Analytical Approach . . . 203
8.4 Results . . . 205
8.4.1 Emerging Hotspot Analysis . . . 206
8.4.2 Spatial Regression . . . 208
8.5 Discussion . . . 210
8.6 Conclusion . . . 213
References . . . 213

9 Urban Agriculture and the Neighborhood Context: A Geodemographic Approach to the City of Toronto . . . 217
9.1 Introduction . . . 217
9.2 About the City of Toronto . . . 219
9.3 Background . . . 221
9.3.1 Green Infrastructure and Urban Processes . . . 221
9.3.2 Practical Considerations: The Need for Analysis . . . 223
9.4 Methodology . . . 224
9.4.1 Data . . . 224
9.4.2 Data Reduction . . . 225
9.4.3 Segmentation . . . 226
9.5 Cluster Descriptions . . . 227
9.6 Analysis of Urban Agriculture Distribution . . . 231
9.7 Discussion . . . 235
9.8 Conclusion . . . 236
References . . . 237

Index . . . 239

Chapter 1
Land-Use and Land Cover Change: Advancing with Geographic Information Science

1.1 Land Use in Ontario

Canada is one of the largest countries in the world concerning its landmass. With around 10,000 km^2 of land, it holds several geographical-analytical obstacles that hinder a systematic approach to understanding its land-use and land cover dynamics (Meyer and Turner II 1994). While several attempts to classify land use and cover in Canada have been carried out, there still lacks a complete spatial and temporal repository for the entire country. Nevertheless, such a dataset would add intrinsic value to respond to issues of economic growth, sustainable development, monitoring of policies, and environmental changes. Given the extent of Canada, however, novel land-use classification approaches are needed that allow such global land-use analytics. This remains a challenge due to the need for consensus between the priorities of the federal and provincial governments. However, budgeting and cost allocation strategies depend primarily on the priority vectors for each province that rely on the development strategies and economic determinants of regional growth (Leo et al. 1998). Within the prioritizing of understanding land use in metropolitan areas, at the provincial level, a thorough and complete knowledge of land use remains a significant challenge.

The ten provincial divisions of Canada show that Ontario corresponds to the province with the largest population of 14.19 million in its province in 2017, having grown by 637,600 inhabitants between 2013 and 2017 (Fig. 1.1).

The shortage of spatially consistent and temporally explicit data at a fine-grained resolution for the entire province makes it a challenge to understand the provinces' land-use/land cover change.

Given the current challenges in rapid urbanization, agricultural abandonment, climate change, sea-level rise, ecosystem loss, and other factors, it is of utmost importance to mobilize spatial information, particularly land-use data well beyond metropolitan areas (Haider et al. 2022). This is, however, not the case in Ontario,

E. Vaz, *Regional and Urban Change and Geographical Information Systems and Science*, Advances in Geographic Information Science,
https://doi.org/10.1007/978-3-031-24731-6_1

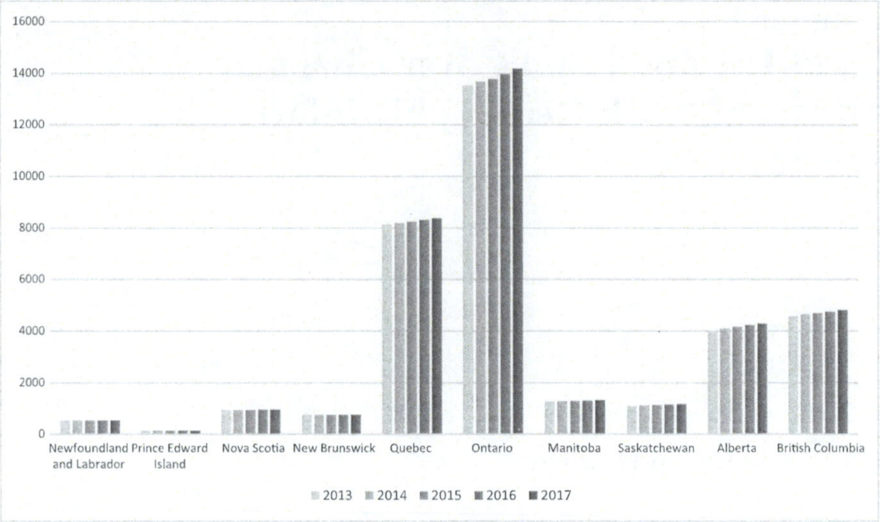

Fig. 1.1 Changes in population between 2013 and 2017

where most of the land-use data available remits to a perimeter confined to a buffer around the most significant urban areas. The key reason is that current high-density population areas in metropolitan regions work as catalysts of economic growth (Vaz and Arsanjani 2015). The detailed mapping of land use/land cover in the regions that must consider transportation efficiency, real estate, commercial activity, and urban planning is often an initiative of privately held corporations. Their investment in highly qualified personnel for surveying and land-use classification delivers a viable commercial product to the province. Within the provincial planning structure, however, this creates a growing asymmetry in the availability of digital information. While land-use/land cover data becomes increasingly available for several timestamps in Ontario, the northern part of the province becomes progressively more isolated in terms of digital information.

Figure 1.2 shows the availability of land-use data through DMTI Spatial for 2014 (Vaz and Arsanjani 2015), compared with the rest of the province. DMTI Spatial is a Canadian market leader in digital mapping with over 20 years of experience, focusing its products on (i) real-time integrated location solutions, (ii) location-enabled predictive analytics, and (iii) location data products.[1]

Of an approximate area of one million square kilometers of the province, only 44,514 km^2 are covered by DMTI. This corresponds to approximately 4.5% of land use/land-use cover for the province, albeit within the urban cores. This corresponds to a small fraction of the regional and provincial challenges that should be examined. The DMTI land-use dataset is readily available for key research institutions in Ontario through licenses with academic libraries through Scholars GeoPortal [1].

[1] https://www.dmtispatial.com/our-story/ (retrieved November 25, 2019).

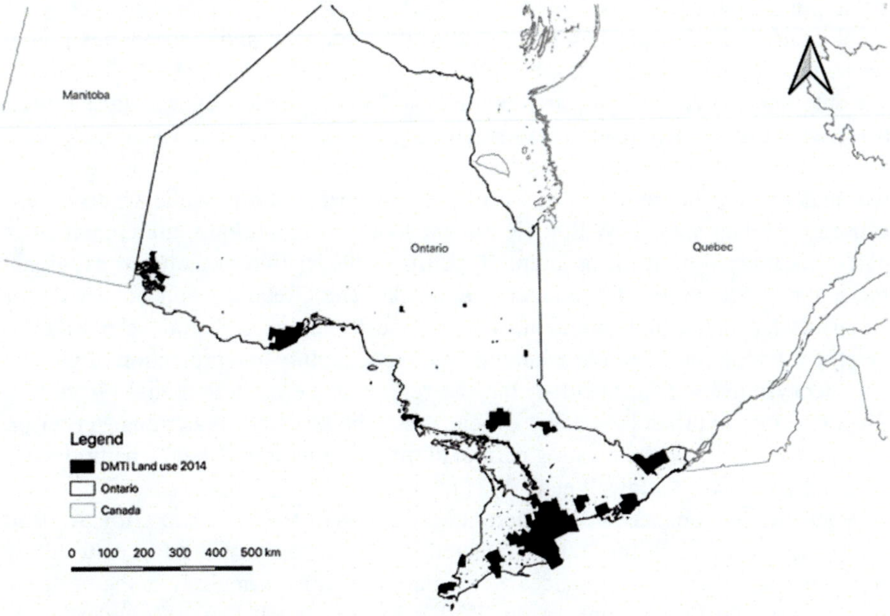

Fig. 1.2 Distribution of land-use classification by DMTI in 2014

This infrastructure holds hardware and software that provides access to large-scale geospatial datasets, as well as sophisticated search, discovery, and analysis tools to make the data readily available to Ontario researchers. This is an excellent combination of private and institutional resources; however, one must consider what geographical research can be carried out by researchers and stakeholders for Northern Ontario given the significant shortage of available digital information.

1.2 Ontario's Greenbelt

The unprecedented urbanization Ontario has witnessed in recent decades concentrates predominantly in Southern Ontario, leading to a stark contrast between Ontario's north and south. This brings discrepancies between policy implementations and strategies within the province. The importance of controlling urbanization beyond a certain extent, for instance, is well nested in the context of Ontario's greenbelt as an attempt to control the impact on environmentally sensitive areas (Ndubisi et al. 1995; Kosmas et al. 2006; Feizizadeh et al. 2023).

Defines preservation areas, while the Places to Grow Act (2006) reports on existing communities within the Greater Golden Horseshoe area where future urbanization may exist. The locational characteristics of both plans aim to respond to the need to create a sustainable regional environment, ensuring socio-economic growth at the provincial level (Macdonald et al. 2021). The Greenbelt Plan integrates geographic information regarding the extent of the Greenbelt and incorporates policy strategies for its proper functioning (Mitchell et al. 2021). As such, its function is not only one of delimiting its area but also accounting for its natural, agricultural, and heritage systems, creating and stabilizing habitat preservation and ecosystem dynamics. The diversified land use/land cover within Ontario's Greenbelt creates a regional hub for sustainable land use, protected from the possible deterioration caused by urban sprawl. With a significant focus on agriculture, the Greenbelt is part of the more significant agricultural industry, offering two specialty crop regions: the Holland Marsh and the Niagara Peninsula. The Greenbelt concerns 6.1% of Ontario's farmland, and interactions with land-use change should be monitored carefully at a regional level to guarantee the sustainability and protection of vulnerable ecosystems and farmland within the region at the highest spatial resolution. The proximity to urban land development should be assessed concerning a criterion of negative impact on the existing natural heritage, agriculture, and the natural environment (Ali 2008; Caldwell et al. 2021).

Figure 1.3 encompasses the different boundaries of the *Places to Grow Act* and the *Greenbelt Plan*. Within the Delineated Built Boundary of the Places to Grow Act, one can see in some instances a boundary overlap, particularly in the northern limits of the Greater Toronto Area. Within the Greenbelt Outer Boundary, it is delimiting the regions' extension of the Greenbelt system.

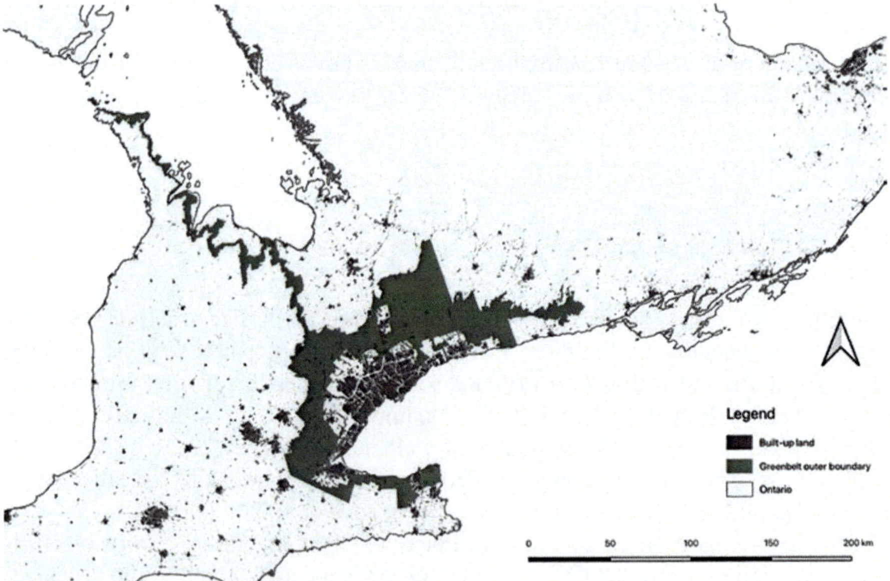

Fig. 1.3 The Greenbelt of Southern Ontario

With the ongoing challenges of urbanization in the outskirts of the Greater Toronto Area, Southern Ontario should carefully monitor the impacts of urban land use. The drastic change in Southern Ontario's land use, brought by the rise of Toronto as an economic force within North America, suggests that careful monitoring of land use and landscape measuring should be considered (Gayler 2009; Cadieux et al. 2013).

Given the large extent of the province, novel approaches for measuring and classifying land use are vital for the efforts of sustainable development. Crowdsourced data may become an essential asset in the future, gathering crucial variables such as subjective well-being, transportation and mobility, environmental perception, and even land-use data at a reasonably high spatial resolution and with substantial spatial accuracy (Liu 2021). Crowdsourced monitoring could significantly reduce the cost of surveying and would allow a technological solution for the more remote areas within Ontario with some challenges related to accuracy assessment and data quality (Yin et al. 2021).

While such a crowdsourced system is not set in place for the province, the current availability of global land cover datasets, such as the GlobeLand30 initiative (Chen et al. 2017; Balogun et al. 2022), abridges well a consensual understanding of land use for the entire province. This dataset in particular with two editions (2000 and 2010) holds as an important monitoring tool for land-use analysis for the entire province of Ontario. The extent of Ontario calls for an integrative approach of digital land-use data with consistent requirements regarding data standards, spatial resolution, and temporal uniformity. Such a dataset would allow a more integrative assessment that could respond to local and regional problems within the province. An excellent example stems from the European CORINE Land Cover (CLC) initiative (Büttner et al. 2004). This program, with its first edition in the mid-1990s, created a digital inventory of land cover for the 27 European Community member states and other European countries, at a scale of 1:100,000 and comprising 44 classes and a three-tier nomenclature. The availability of seamless vector data for five distinct timestamps (1990, 2000, 2006, 2012, and 2018) creates a unique instrument for spatial and regional analysis of land-use-related issues in Europe. Such a joint initiative would propel the understanding of regional land-use challenges in the province, particularly when relating to the integration of spatial analysis and Geographic Information Systems to explore, analyze, visualize, and assess provincial problems at micro- and macro-level of policy interaction (Feranec et al. 2007). The rural hinterlands of South Ontario and North Ontario, in general, are only sparsely classified regarding land use and land cover. It should be noted that the absence of research dealing with land use and land cover in Ontario limits the potential to readjust global provincial policies and guarantee an integrated approach for global sustainable development of the province considerably.

A closer inspection of the publication records found in Scopus (Fig. 1.4) allowed us to assert a growing number of research interests since the early 2000s, averaging since 2000 around 36 publications per year. This compared with the average of seven publications per year on this research topic between 1955 and 2000. The considered source types pertain to 812 journal publications, 59 conference proceedings,

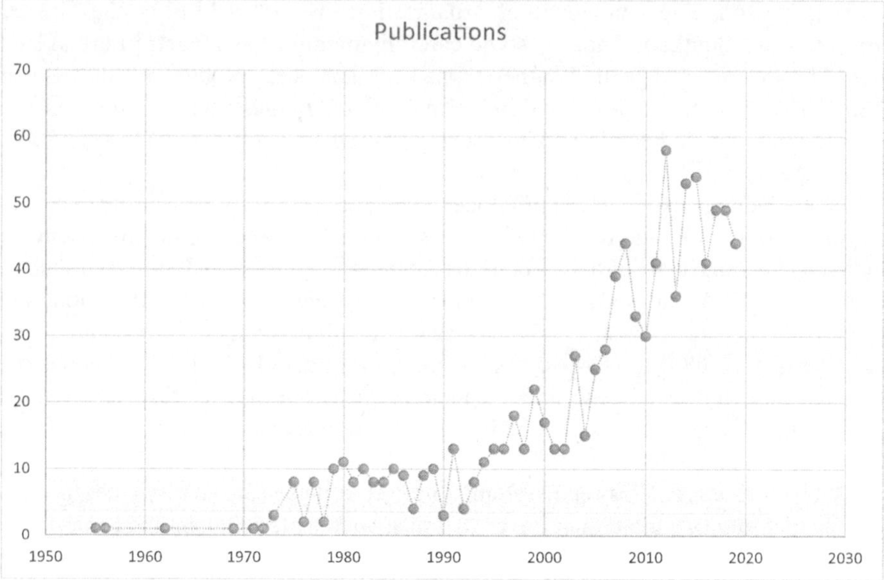

Fig. 1.4 Increase in publications dealing with land use in Ontario

35 books, 29 trade publications, and 20 book series. Ninety-eight percent of these publications are in English, two in French, and one in Chinese and Japanese, respectively. Sixteen of these publications had undefined languages. Given the bilingual nature of Canada, a second metadata search was performed with the query "Ontario" AND "utilisation des Terres," without any significant results to the search. The five most research-intensive countries exploring since 1955 the topic of land use in Ontario are Canada, the United States, the United Kingdom, China, and Sweden.

Additionally, an inspection between research on the topic between Canada and the United States between 2005 and 2019 allows to conclude that patterns of research interest for this topic are similar with both countries registering a peak in 2012. A simple ANOVA regression allowed to predict a point prediction of circa 50 publications per annum in 2030.

The availability of high spatial resolution land-use data outside the metropolitan areas of Ontario would certainly enhance a research framework given the many publications dealing with the environmental sustainability of the province. In this sense, the lack of datasets for North Ontario brings a loss of relevant information structures that could enhance a research and policy agenda and lead to potential innovation in regional planning structures. Additionally, the complete reference for Ontario's land use can be found at a spatial resolution of 30 m. The LU (Land Use) project covers all areas of Canada south of 60°N and pertains to the following land-use classes: unclassified (due to cloud coverage), settlement (built-up and urban), roads (primary, secondary, and tertiary), water (natural and human-made water sources), forest (treed areas greater than 1 ha in size), forest wetland (wetland with forest cover), trees (treed area less than 1 ha), treed wetland (wetland with tree cover), cropland (annual and perennial crops), grassland managed (natural grass and shrubs used for cattle grazing), grassland unmanaged (natural grass and shrubs with no apparent use), wetland (undifferentiated wetland), wetland shrub (wetland with shrub cover), wetland herb (wetland with grass cover), and other land (rock, beaches, ice, barren land). Given the availability of several time stamps, a systematic analysis can lead to a better understanding of land-use change within the province, despite its spatial resolution and absence of subclass systems for understanding land-use transitions in significant detail to positively contribute to provincial decision-making. This regional understanding is relevant to guarantee the correct land management at present due to endogenous and exogenous factors.

1.3 Ontario's Land-Use Organizations

A historical overview of land use and land cover is fundamental for sustainable development. Within the province of Ontario, such an initiative is linked to the Ministry of Natural Resources and Forestry (MNRF) genesis. The complexity, however, related to the existence of different data sources, local projects, and the intrinsic complexity of intraorganizational communication has not allowed completing a global picture of the various endeavors that have been contributing over the last decades. Such a contribution would have a paramount value to the community and stakeholders. This results from spatial-explicit methodologies becoming increasingly essential tools for efficient regional decision supports in progressively complex land-use interactions in the province.

To this adds the historical legacy of key contributors to the field of Geographic Information Science in Ontario by means of founders of the discipline, such as Michael Goodchild and Roger Tomlinson. One should not forget the pioneering role of the province in the mid-1970s as stakeholders in introducing the integration of Geographic Information Systems into provincial decision-making and computation as a tool for better administrative regional management. The adaptation of Geographic Information Systems into administrative and regional decision-making and the innovation shaped by key stakeholders in proposing computational solutions paved the roads for geocomputation and spatial data in general.

The legacy and preservation of this key institutional knowledge and path dependency should be reassessed in the province's abundance and availability of spatial data. There is a pressing need for advancing significant research in Geographic Information Science that requires complete and robust data to address multidisciplinary issues within land use, spatial data science, and regional science in general.

The history of the land-use/land cover tradition in Ontario is set by the Lands and Waters Group, established in 1977, within a central role of the conservation authorities' branch, recognizing the importance and the importance of understanding the different changes within Ontario's land-use structures. This group, in effect between 1977 and 1990, created several agencies to define the complexity of remote sensed initiatives such as (i) the Land Use Co-ordination Branch (1977–1983), (ii) the Surveys and Mapping Branch (1977–1979), the Engineering Services Branch (1977–1983), the Conservation Authorities Branch (1977–1979), the Conservation Authorities and Water Management Branch (1979–1990), the Land Management Branch (1979–1990), the Office of Indian Resource Policy (1977–1990), and the Ontario Centre for Remote Sensing (1985). It was the Ontario Centre for Remote Sensing (OCRS) located in Toronto that constituted an essential part of the Ministry of Natural Resources, responsible for conducting applied research projects and thematic mapping programs within the fields of forestry, geology, and land-use planning, utilizing the assessment of airborne and spaceborne remote sensed imagery. This allowed, together with the recently available Landsat TM imagery, to encompass the creation of the first Provincial Land Cover dataset, and the rise of a new era of land-use inventory for the province. The dates between 1986 and 1997 were recorded in synergy with the OCRS work on advancing with applied research in the field, leading to a comprehensive dataset becoming available in the early 1990s. The success of these first land cover sets was largely a result of the adequacy of classification, with the lowest classification accuracy of 85% along with the agricultural land cover, also resulting from the existence of the many small towns and remote road networks that led to misclassification. Overall, the classification accuracy suggested the first benchmark on a first vision of the complexity of land-use classification in Ontario. Despite the defuncting of the OCRS, the availability of the Provincial Land Cover in its second edition allowed for continuous interest and production of land-use and land cover data passing along the tides of different governmental restructuring. Thus, the legacy of Ontario's land cover and land-use classification is strongly linked in its genesis to the appearance of available global land-use imagery, and pioneering endeavors brought forward by the Ministry. The current fragmentation of available datasets, however, is a result of a lack of a cohesive effort, which denotes changes in internal dynamics, where over the last 30 years, much of the results are still today available in the resulting publications dealing with land-use change and impacts of forest and water bodies as well as agricultural land within the province. This research suggests an intrinsic interest in understanding the province at a regional level, where despite the difficulty in harnessing a contextual understanding of the evolution of the tradition of land-use inventories in the province of Ontario, it becomes clear that it is the effort in the 1990s, through several individuals that pioneered in understanding the physical and human impacts of land-use change, that created a first vision of the usability of these datasets to understand Ontario's regional challenges in the future.

1.4 The Importance of Regional Models for Land Use

A plethora of different usages of land-use models are available at the regional level. The computational limitation of handling global-scale data while keeping a higher resolution leads to a more abundant number of regional land-use models. These models address issues such as climate change, ecosystem challenges, agricultural land transitions, and urban sprawl, to be addressed, usually at a regional level. However, the growing advances in the last decade of computational power (chiefly due to faster processors and RAM integration) and open-source tools to tackle land-use models have redefined the role of the regional analysis.

The geographical boundaries of analysis have become larger with the increase of computational power, and due to open and often crowd-based data, the capacity of rigorous classification algorithms has become better. This has brought a more profound understanding in regard to land-use composition and its effects on the sustainability of ecosystems and society. The advent of landscape metrics, for instance, has introduced multidisciplinary quantitative methods to assess regional land-use change, creating a deterministic understanding of land-use change and its impacts at the provincial level. It is fair to say that if it were not for geocomputational advances, only a restricted geographical boundary would be feasible to compute at such a fine spatial granularity. The computational advance and availability of novel algorithms and spatial analysis methods have significantly contributed to assess Geographic Information Science in the province of Ontario.

At this point, a difference between static and dynamic land-use models should be noted. Geographic analysis, in particular when addressing issues of land-use/land cover change, addresses mostly interdisciplinary research endeavors that can benefit from quantitative approaches. Such quantitative models take into account the spatial resolution of available land-use data (chiefly for static land use interpretation), or past and present land-use coverage for tracking changes. The range of modeling approaches is thus dependent on (i) data availability, (ii) resolution of land-use data, (iii) research objective, and (iv) temporal stamps (depending on the interest of investing in static or dynamic spatial models). These then resort to deductive or inductive approaches to lead to a thorough understanding of land use that frequently are quantified utilizing descriptive statistics or advance toward nonlinear modeling approaches to predict changes and possible driving factors of land-use change. These underlying factors can then be integrated within socio-economic variables and determine land-use capacity, bringing an objective understanding of patterns of change merging at provincial-level regional science with geography as well as other scientific scopes.

1.5 Governmental Initiatives of Land-Use Collection

Synergies between different organizations allowed for a creation of joint endeavors addressing main concerns of land use such as water management, agriculture, urbanization, and forestry. These enabled a land-use classification inventory from the 1960s to 1995. The existence of different levels and tiers of land-use information throughout the Canada Land Inventory (CLI) resulted in such an effort, bringing a cohesive knowledge of Canada's land use for over three decades. The CLI was carried out by a combination of aerial photography, existing cartography, and ancillary metadata. Within its core objective, as is in the tradition of Canadian land-use exploration, the CLI aimed to integrate a concise understanding of Canadian agricultural suitability, focusing on crop types and production. The approach brought seven distinct classes adequate for supervising and assessing agricultural activity.

Frequently, however, land-use data is harnessed from a set of different open data sources, which nevertheless have an impact on the granularity and different data standards, leading to discrepancies of land-use classification and land-use interpretation. The different scales from several projects are an essential characteristic when choosing land use. When tracing the changes in metropolitan areas, mainly due to rapid urbanization and expansion, it is vital to consider the urban land-use morphology concerning its density and sprawl, as to understand its impact and characteristics throughout the spatiotemporal extent of analysis. In the case of Ontario, particularly Southern Ontario, a significant amount of urban land-use change has been witnessed, leading to previously compact urban regions accruing additional urban sprawl in recent decades. This is directly in the case of Southern Ontario as a result of the inflow of migration and socio-economic growth, followed by the allocation of real estate and the significant growth of the housing markets. Within the Greater Golden Horseshoe region, the Greater Toronto Area has been particularly affected due to its rapid economic growth, and its performance as one of the largest economies in the North American market. In this sense, one must consider that urban regions inevitably will become part of Ontario's landscape. The dynamics of urban growth as well as the verticalization of cities such as Toronto lead to population increase that is expected to continue to grow in the next decades. Regions must rapidly transform their infrastructure as to cope with population increase and accommodate socio-economic as well as commercial growth without jeopardizing land use and land cover. The compactness of urban regions becomes thus a resoundingly important topic of the future of land-use sustainability, where the conditioning of land-use change must be thought in regard to the impact on agricultural, natural, and fragile ecological regions. Southern Ontario and the province must consider the increase of 72% of the population expected to be living by 2050 in urban areas. The Greater Toronto Area itself sets out as the ideal laboratory of urban change in the coming decades, where Canada can be a leader of sustainable urban growth. Toronto already holds one of the most abundant forest canopies in the world. Despite several challenges relating to urban sprawl in the northern perimeter of the Greater Toronto Area, there is at present a significant opportunity to incorporate sustainable planning and land use in the region.

The existence of the Greenbelt and the mitigation strategies of the Places to Grow Act are all part of an ensemble of a sustainable vision of Southern Ontario's sustainable urbanization. Balancing the existing asymmetry between Northern and Southern Ontario remains a significant challenge, as the bulk of Ontario's economy is concentrated in the south, while the north boasts a wealth of resources and small communities that need to be included in the drive towards equal growth across the province. Only be offering an integrated strategy at the provincial level for the dichotomy between Ontario's north and south can the province itself claim a balanced and equitable vision of sustainable growth. Land use and land-use change are of the utmost importance in this transaction and play a leading role in the future of sustainable urbanization in the province.

1.6 Urban Land-Use Planning in Ontario

In recent decades, a substantial amount of change has been witnessed in Ontario's land-use regulation. With the incorporation of the Greenbelt Plan, the Places to Grow Act allocation of land use has gained a relevant policy framework. The purposes of the Act are as follows:[2]

(a) To enable decisions about growth to be made in ways that sustain a robust economy, build strong communities, and promote a healthy environment and a culture of conservation
(b) To promote a rational and balanced approach to decisions about growth that builds on community priorities, strengths, and opportunities and makes efficient use of infrastructure
(c) To enable planning for growth in a manner that reflects a broad geographical perspective and is integrated across natural and municipal boundaries
(d) To ensure that a long-term vision and long-term goals guide decision-making about growth and provide for the coordination of growth policies among all levels of government

The Environmental Commissioner of Ontario (ECO) has been in the forefront of regulating laws that constitute adequate interaction of policy and governance, such as (i) the recommendation of Ontario Municipal Board, (ii) the different regional plans within the province, and (iii) the planning structures for growth and transportation. These have been particularly relevant topics for the decision support of Ontario's present and future land use. The focus of the Planning Act brings essential aspects of environmental assessment and management, where the key priority is the diversity of ecosystems and sustainability of agricultural land while offering an optimized vision of urbanization without compromising environmental resources and creating additional waste. The Planning Act states clearly its purpose: "promote sustainable economic development in a healthy natural environment within a provincial policy framework."[3] It is important to note that the Planning Act has to keep a sustainable landscape following criteria of preservation and diversity of land use and to communicate with the province and regions the relevant measures to warrant sustainable development.

In this sense, the legislation offers a normative framework concerning the use and control of privately owned lands through the incorporation of available planning tools. Additionally, the Planning Act also fuels the engine of planning systems

[2] https://www.ontario.ca/laws/statute/05p13

[3] https://www.ontario.ca/document/citizens-guide-land-use-planning/planning-act

that allow integrating concerns of the province within the local decisions of municipal stakeholders. Transparency and consistency between the Provincial Policy Statement and the Planning Act should thus lead to an authoritative synergy. Such a synergy should drive a rich and multidimensional dialogue between municipal councils, local and regional planning boards, provincial ministers, and the provincial government, where agency officials become a part that further negotiates with the Ontario Municipal Board. Prioritized by the Planning Act, the primary objective constitutes an inability to counteract under any circumstance the existing provincial plans, such as the Greenbelt Plan, guaranteeing the protection of vulnerable and protected regions within the province. In this sense, it is relevant to consider the role of urban expansion efficiently, to avoid encroachment and potential land-use fragmentation as the southernmost part of the province struggles with population increase. Monitoring the efficiency and efficacy of land-use/land cover change is thus of paramount importance to reassess current impacts on fragile ecosystems. An example of this is the unavailability of data that examines smaller land-use change within wetland systems that are often unreported despite some existence of data. A consequence of the lack of monitorization of small areas throughout the province leads to a series of constraints on sustainability of planned transportation networks and the resulting urbanization which should consider the interaction between economic growth and environmental sustainability as well as transportation efficacy.

While we have seen that the initiatives for regional land-use classification in Ontario have been strongly linked to the exploration of governmental initiatives, leading to a robust framework of a land-use/land cover analysis culture, it is important to position the debate in the current alternatives to following in situ classification, having in account the size of the province of Ontario. Reliable land-use data, at high resolution and throughout such a large land-use body, is generally difficult to organize and compile and may well be in the genesis of the lack of completion of Ontario's land-use standardization within land cover and land-use classification over the decades. Furthermore, Ontario's urban environments have specificities that relate to subtypes of land-use categorization, where at a provincial level, similar to the loss of smaller towns and settlements, due to the dimension of the sheer land-use mass itself, it becomes difficult to create a significant disaggregation that allows distinction of subcategories of economic activity related to land use. The increasing proximity of the Golden Horseshoe, for instance, of urban land, where rural land depletion is eminent, is such a concern, fostered by urbanization from the Greater Toronto Area. As noted by 5.80% of agricultural land and 5.47% of rangeland were lost between 2000 and 2010, as a consequence of the urban sprawl on Toronto's fringe. Historically, the mandate of Environment Canada has brought a tradition to understand and tackle the challenges of land-use change.

Frameworks such as the Canada Land Use Monitoring Program (CLUMP) aim to generate awareness of the current monitoring of land-use types while mitigating land-use conflicts and adhering to the rigorous assessment of land-use change. This landscape of understanding land use in Ontario also has a new opportunity within Geographic Information Science. The advance in volunteered geographic information (VGI) has changed the availability of land-use data through a bottom-up approach. Considering the increasing availability of VGI data such as the one provided by OpenStreetMap (OSM), with enough users, a more in-depth understanding of Ontario's land use may become available. These datasets create unique integrative strategies that may enable us to measure and assess land use/land cover in detail and spatial extent that previously was impossible.

1.7 Volunteered Geographic Information as Land-Use Sources

Despite the international achievements of producing land use and land cover for regional assessments, often such initiatives have a coarse resolution leading to difficulty in obtaining a local assessment. The growing approaches that combine available global land coverage with high-resolution imagery are thus an essential tool for adequate land-use classification in Ontario. With the advances of OSM, land-use classification gains a new aspect of thorough land-use classification, where local authorities may gear open data toward goals important for public policy. Crowdsourced data has the intrinsic property of multi-scalability, which eases the otherwise cost-inefficient and time-consuming process of surveying that for large terrestrial surfaces is quite a challenge. Such technologies recognize the accuracy and feasibility of gathering, editing, and joining spatial information through handheld devices, enabling a local classification of land-use types and land cover distribution. Several web map services, such as ESRI's Base Maps, Bing Maps, and Google Maps, have been at the forefront of allowing application programming interfaces for end users. These frameworks interact with high-resolution spatial products, particularly when nested with volunteered geographic information repositories such as OpenStreetMap. In this sense, collaborative mapping projects have brought locational positioning for land-use information that is generated from users. This harnesses a novel type of spatial data, enriched at a very high resolution, and with the potential of becoming a multiuser collaborative platform to design and update the accuracy of land use/land cover.

The characteristics of volunteered geographic information (VGI) bring a unique opportunity for quality classification of land use that remains to be explored for Canadian provinces. OpenStreetMap (OSM) stands out as one of the most vital resources for a collaborative mapping environment. The potential of downloading consistent spatial data and the frequent additions to the refinement and update of datasets leads to the possibility of creating (i) rigorous, (ii) temporally complete, and (iii) regionally significant datasets that otherwise would be difficult to achieve. Investigates the higher accuracy brought forward by such data sources regarding the importance of classifying urban land use compared to traditional data sources. The usage of OSM, however, and thus in the larger sense, the contribution to VGI for land-use classification for more significant regions, is still in its embryonic stages. Nevertheless, it has been proven that in developed countries, the consistency of land-use classification not only is available but also permits a direct relation to areas that suggest regional disparity. In the case of Ontario, this is of extreme importance, given the difference between Southern Ontario and Northern Ontario, as well as the tremendous difference of economic activity throughout the province.

1.8 Conclusions and Future Work

VGI is a promising endeavor for the future land-use classification of growing urban regions in the developed world. The advances in contributing technologies and the increased population density within larger metropolitan regions suggest that collaborative projects, such as OSM, can support the land-use planning agenda. In the case of Canada, where land-use classification poses great logistical challenges due in part to issues of regional administration and jurisdiction, and cost efficiency, the integration of land-use classification through VGI is a very welcome approach for more accurate land-use information. VGI-based approaches for land mapping have been further tested by Geo-Wiki (Fritz et al. 2009) for global coverage, and promising results for VGI-based mapping of land features were achieved.

The spatiotemporal result of land-use classification that becomes available through this integrative approach further allows the sharing of information that portrays spatial dynamics vis-à-vis land-use change in Canada, one of the most important phenomena to consider in rapidly changing urban regions. Toronto, which is on the verge of becoming a megacity in the coming decades, needs public information and expert knowledge to cope with a sustainable urban future in order to mitigate the risks for Ontario's greenbelt and offer adequate monitoring of the urban sprawl. The collective force brought by a Web 2.0 context in lieu of traditional surveying methods suggests a very promising approach to measuring, monitoring, and assisting land-use changes and transitions in rapidly changing environments.

References

Ali AK (2008) Greenbelts to contain urban growth in Ontario, Canada: promises and prospects. Plann Pract Res 23(4):533–548

Balogun AL, Mohd Said SA, Sholagberu AT, Aina YA, Althuwaynee OF, Aydda A (2022) Assessing the suitability of GlobeLand30 for land cover mapping and sustainable development in Malaysia using error matrix and unbiased area estimation. Geocarto Int 37(6):1607–1627

Büttner G, Feranec J, Jaffrain G, Mari L, Maucha G, Soukup T (2004) The CORINE land cover 2000 project. EARSeL eProceedings 3(3):331–346

Cadieux KV, Taylor LE, Bunce MF (2013) Landscape ideology in the Greater Golden Horseshoe Greenbelt Plan: negotiating material landscapes and abstract ideals in the city's countryside. J Rural Stud 32:307–319

Caldwell W, Geschiere E, Sousa E, Zink R (2021) Municipal capacity: a case study of Ontario's greenbelt to respond to emerging agriculture and agri-food priorities. Int J Environ Impacts 4(3):243–261

Chen J, Cao X, Peng S, Ren H (2017) Analysis and applications of GlobeLand30: a review. ISPRS Int J Geo Inf 6(8):230

Feizizadeh B, Omarzadeh D, Kazemi Garajeh M, Lakes T, Blaschke T (2023) Machine learning data-driven approaches for land use/cover mapping and trend analysis using Google Earth Engine. J Environ Plan Manag 66(3):665–697

Feranec J, Hazeu G, Christensen S, Jaffrain G (2007) Corine land cover change detection in Europe (case studies of The Netherlands and Slovakia). Land Use Policy 24(1):234–247

Gayler HJ (2009) Ontario's Greenbelt and places to grow legislation: impacts on the future of the countryside and the rural economy. In: The next rural economies: constructing rural place in global economies. CABI, Wallingford, pp 75–88

Haider A, Rankaduwa W, ul Husnain MI, Shaheen F (2022) Nexus between agricultural land use, economic growth and N2O emissions in Canada: is there an environmental Kuznets curve? Sustainability 14(14):8806

Kosmas C, Tsara M, Moustakas N, Kosma D, Yassoglou N (2006) Environmentally sensitive areas and indicators of desertification. In: Desertification in the Mediterranean region. A security issue. Springer, Dordrecht, pp 525–547

Leo C, Beavis MA, Carver A, Turner R (1998) Is urban sprawl back on the political agenda? Local growth control, regional growth management, and politics. Urban Aff Rev 34(2):179–211

Liu Z (2021) Identifying urban land use social functional units: a case study using OSM data. Int J Digital Earth 14(12):1798–1817

Macdonald S, Monstadt J, Friendly A (2021) Towards smart regional growth: institutional complexities and the regional governance of Southern Ontario's Greenbelt. Territory Politics Gov:1–21

Meyer WB, Turner BL II (eds) (1994) Changes in land use and land cover: a global perspective, vol 4. Cambridge University Press, Cambridge

Mitchell B, Shrubsole D, Watson N (2021) Ontario conservation authorities–end, evolve, interlude or epiphany? Can Water Resour J/Revue canadienne des ressources hydriques 46(3):139–152

Ndubisi F, DeMeo T, Ditto ND (1995) Environmentally sensitive areas: a template for developing greenway corridors. Landsc Urban Plan 33(1–3):159–177

Vaz E, Arsanjani JJ (2015) Predicting urban growth of the greater Toronto area-coupling a markov cellular automata with document meta-analysis. J Environ Inf 25(2):71–80

Vaz E, Jokar Arsanjani J (2015) Crowdsourced mapping of land use in urban dense environments: an assessment of Toronto. Can Geogr/Le Géographe canadien 59(2):246–255

Yin J, Dong J, Hamm NA, Li Z, Wang J, Xing H, Fu P (2021) Integrating remote sensing and geospatial big data for urban land use mapping: a review. Int J Appl Earth Obs Geoinf 103:102514

Chapter 2
Urban Sprawl into the Niagara Region: Urban Encroachment on Agriculture

2.1 Introduction

The Niagara Region is an area made up of pristine ecosystems and prime farmland. Even though the land is so valuable, development projects are prominent. The study area involved in this analysis incorporates a small portion of the Niagara Escarpment, and therefore, it is important to understand the laws and regulations that surround the development along the Escarpment area. The Niagara Escarpment is mostly a protected zone, and yet housing expansions and industry/commercial developments are dotting the landscape. Urban sprawl occurs when larger cities continue to grow. It is defined as low-density expansion of larger urban areas that manifest in a physical pattern due to certain market conditions. This expansion is often found to overtake some agricultural spaces. Sprawl is dangerous as it is synonymous to development which is not well planned, or planned at all, and development which occurs incrementally (EEA 2006). Once the population can no longer be contained within the city limits, housing on the outskirts of the city center must be created. This development expands the city limits and the pattern continues outward if there is no policy and management in place. As city centers grow, they need building materials which are sourced from the surrounding rural areas, or shipped in from elsewhere. Not only do cities expand outward in the physical sense, but the growth of a city can be seen in the boom of industry in the surrounding rural areas which supply many resources. To observe whether or not urban sprawl is having any effect on the Niagara Region, landscape metrics will be utilized. Spatial analysis involving landscape metrics in ArcGIS will disclose whether there have been significant changes over time in agricultural land. Remote sensing software will allow for land use to be shown using specific land-use classifications. Land-use change will be shown over time. The dates chosen span a 30-year period from 1985 to 2015. This ensures that change detection can be completed by comparing and

E. Vaz, *Regional and Urban Change and Geographical Information Systems and Science*, Advances in Geographic Information Science,
https://doi.org/10.1007/978-3-031-24731-6_2

contrasting satellite imagery from those years. The following will be a comprehensive literature review that includes aspects of policy, geologic composition, and settlement in the Region.

The urban-rural fringe is a particularly sensitive zone for agriculture. Exurbia, or the urban-rural fringe, is one of the fastest changing landscapes today (Shaker and Ehlinger 2007). As cities push out of their boundaries, agricultural land is developed into low-density residential neighborhoods. This poses a problem called fragmentation. While fragmentation can occur naturally, most research as of late discusses human-imposed fragmentation. Fragmentation along the urban-rural fringe is an important consideration in planning processes (Petrov and Sugumaran 2009). The research in this thesis focuses on the loss of agricultural lands throughout the Niagara Region with the expansion of many urban populations. The Niagara Region was chosen for the study due to the area's distinct agricultural possibilities. The territory has some of the best farmland in all of Ontario, and urban sprawl is encroaching, which has and will continue to cause many issues. These issues include the size of farmland, fragmentation, soil erosion, and the loss of agricultural zones entirely.

2.2 Study Area

Farmland loss needs to be monitored for a number of reasons. Farmland has an effect on the economy of the Region, it varies the landscape providing enjoyment for those that live in rural communities and for those touring the Niagara Region and Escarpment, and it provides habitat for a number of species, including birds, fish, and other wildlife. Perhaps one of the most important aspects of farmland is that it provides food for human consumption. Due to the importance of these issues to the environment and humans, this study will determine if urban sprawl is affecting agriculture in the Niagara Region. The population of the Regional Municipality of Niagara as of 2011 was 431,346. This is up from 403,554 in 1996 and continues to climb steadily with the advent of new housing plots (Niagara Region 2011). The study area extends north to south from Lake Ontario to Lake Erie and east to west from Niagara Falls to Hamilton, but does not include Hamilton. This is effectively known as the Niagara Peninsula, or the Regional Municipality of Niagara. The study area is portrayed in Fig. 2.1. There are 12 municipalities within the region. From west to east, these include Grimsby, West Lincoln, Wainfleet, Lincoln, Pelham, St. Catharines, Thorold, Welland, Port Colborne, Niagara-on-the-Lake, Niagara Falls, and Fort Erie.

Though Niagara Falls saw the greatest increase in residents by 2016, the actual growth was much larger in Grimsby. Niagara Falls showed a growth in those 20 years of 14.50%, but Grimsby showed an incredible increase of 39.46%. West Lincoln and Lincoln townships are not that far behind with the increased growth trend, showing both 25.94% and 26.52%, respectively. It could be conceivable to state that this is due to these municipalities' spatial approximation to Hamilton, which is a rather large census metropolitan area (CMA) with approximately three

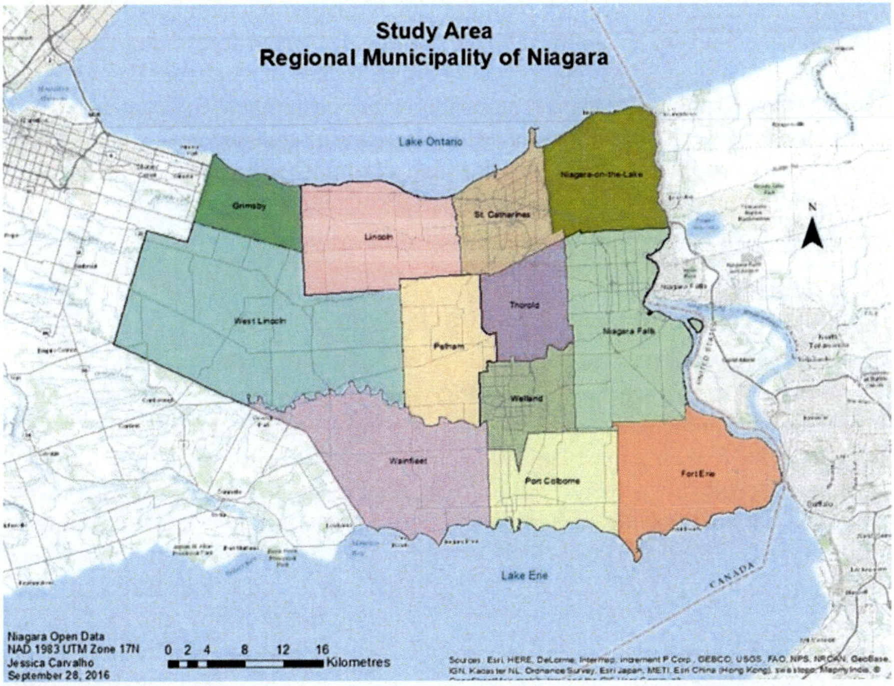

Fig. 2.1 Study area

quarters of a million residents. Another reason for the growth could be the proposed expansion of Metrolinx GO train and bus services. The expansion from Burlington straight through to Niagara Falls for their train services is expected to be completed by 2023, but bus routes run every hour every day. The expansion of public transport makes it easier for those working in a CMA to commute, which means that the less expensive housing outside of those CMAs will likely be more popular. St. Catharines is the largest city in the Regional Municipality of Niagara, but growth is slow at 1.67% change in 20 years. It is crucial to understand why the Niagara Region and, contained therein, Niagara Escarpment, are such unique areas. This section outlines the historic constituents that took place in order for the land to end up the way that is has, creating such prime agricultural lands. Escarpment is a term used to describe a ridge composed of gently dipping rock strata with a long, gradual slope on the one side and a relatively steep cliff on the other side (Park Planning Branch 1976). The Niagara Escarpment extends from Queenston on the Niagara River to the islands off of Tobermory on the Bruce Peninsula in Ontario (Ontario's Niagara Escarpment 2015a). There is no other land like the Niagara Escarpment throughout all of Canada. It is geologically and ecologically unique, plus it boasts many natural rivers and streams that provide some of the best recreational opportunities in all of Ontario (MNRaF 2014). Ontario's Southern Woodland Natives began experimenting with horticulture around 700 AD. This began a new, sedentary way of life. The day of nomads in what is now Southern Ontario slowly phased out. Iroquois Natives that lived near the Niagara Escarpment began burning patches of forest in order to grow corn. By 1350, beans and squash were being cultivated. These were known as the Three Sisters Crops due to their relationship with one another. The corn stalks provided support for the growing beans, while the squash remained near the soil, effectively preventing weeds from growing and retaining soil moisture. Beans also provide nitrogen for the soil, which acts as a fertilizer (Dickason 2010).

2.3 Protection Groups and Policies

It is important to note what environmentally significant areas are. ESAs, sometimes known as environmentally sensitive areas, are areas of agricultural land that need protection due to their unique landscape or wildlife. The historical value of this land often plays a role in whether or not it is classified as an environmentally significant area. Throughout the Niagara Region and especially the Niagara Escarpment, there are numerous ESAs indicated, with varying levels of protection. In Furberg and Ban's (2012) report of the Greater Toronto Area, they found that in 1985 and 1995, there was a significant overlap between sprawling urban zones and ESAs. 1985 showed a 2.5% overlap and 1995 showed a 6.1% overlap. Between those 10 years, there was an increase of 3.6% overlap. The period between 1995 and 2005, however, only showed an increase of 0.3% overlap. This makes sense as protection of these areas has become a much more recent issue. The protection levels within the confines of the Escarpment are much higher, but this region also offers many valuable resources, and so, sometimes, protection can be considered lax. Even though these ESAs are generally well protected, urban expansion can occur around their borders. This effectively disassociates one ESA from another, meaning that ecosystems may be disrupted by the urban settings between two ESAs.

2.3.1 Niagara Escarpment Biosphere Reserve

A United Nations Educational, Scientific and Cultural Organization (UNESCO) biosphere reserve is created by first recognizing that the residents in the area want this area to be considered a biosphere reserve. That decision is then ratified by a national committee, and finally designated by UNESCO. The purpose of a biosphere reserve is to help humans live and work in conjunction with nature (Parks Canada 2005). In February of 1990, UNESCO and the Man and the Biosphere Programme (MaB) named the Niagara Escarpment as one of the world's biosphere reserves (MNRaF 2014). MaB is an intergovernmental scientific plan launched in 1971 that urges better relationships between people and their environments. It uses people from a wide variety of backgrounds, such as the natural and social sciences, economics, and education, to provide innovative ways of increasing economic growth, but in a manner that is both culturally and environmentally sustainable (UNESCO 2014). There are a total of 16 biosphere reserves in Canada and the Niagara Escarpment is included on that list (The Canadian Commission for UNESCO n.d.). The protected zone covers approximately 183,000 ha stretching from Niagara Falls to just outside of the Bruce Peninsula. This is a 725-km stretch of land that changes drastically throughout. Valleys, hills, waterfalls, and caves are just some of the intricate details that can be found here. The Niagara Escarpment houses over 300 bird species, 39 species of reptiles and amphibians, 53 species of mammals, 90 fish species, and approximately 1800 plant species that include some only found in Canada, such as specific orchid flower varieties (Foster 2000). Not only is it home to this wide variety of life, but it caters to diverse human livelihoods, mostly in the form of agriculture. Others include the tourism industry, forestry, and mineral resource mining. The Niagara Escarpment Biosphere Reserve instituted by UNESCO is financed provincially, but typically run municipally. This means that organizations, such as the Niagara Escarpment Commission (NEC), whose purpose is to ensure that the regulations set forth in the Niagara Escarpment Plan (NEP) are adhered to, must work with each municipality separately and come to a consensus on all issues (Foster 2000). This also means that the Niagara Escarpment Commission is subject to provincial cutbacks, which will be discussed in the Niagara Escarpment Commission section of this study.

2.3.2 *Niagara Escarpment Planning and Development Act (NEPDA)*

The Niagara Escarpment Planning and Development Act was approved by the Ontario Legislature in June of 1973 and has been revised as recently as 2012. The purpose of the NEPDA is "to provide for the maintenance of the Niagara Escarpment and land in its vicinity substantially as a continuous natural environment and to ensure only such development occurs as is compatible with that natural environment" (NEPDA 2012).

2.3.3 *Niagara Escarpment Commission (NEC)*

The Niagara Escarpment Commission is composed of 17 members who are appointed by the Lieutenant Governor in Council. The first nine members are people who have been appointed to represent the public. The remaining eight members are chosen by each municipality making a list of people they think would be suitable. One member is then picked from each compiled list. Those on the lists must be an employee of the municipal government (Ontario's Niagara Escarpment 2015b). Many people who live within the boundaries of the Niagara Escarpment are unaware of the level of protection that the area faces. It is understood that protection exists and many community-based protection groups believe that this protection is full proof. This means that they believe that development is not allowed in any sense and this often leads to misunderstandings and local backlash when a proposal for development has been accepted. Many people fail to realize that there are limiting factors affecting the Commission's ability to enforce proper policies. These include funding cutbacks from the provincial government. Cutbacks have made it extremely difficult for the Commission to respond to questionable development proposals and to initiate educational opportunities. There have also been a number of people on the Commission that do not support the protection of the Escarpment as strongly as some of the other environmentally forward political commissioners. They have spoken outwardly about the need to expand the area for economic purposes; thus, many proposals often make it through the cracks of protection (Preston 2001). This is why local governments and citizens must also play a vital role in ensuring the protection of this ecologically important area.

2.3.4 Niagara Escarpment Plan (NEP)

The Niagara Escarpment Plan is used to indicate how land within the Plan boundaries can be used and managed. The Plan also contains information on how development can proceed within the area. Policies are outlined for the Niagara Escarpment Parks and Open Space System as well. It is set by the Ministry of Natural Resources and Forestry and enforced by the Niagara Escarpment Commission. This Plan was established from the NEPDA as a "framework of objectives and policies to strike a balance between development, preservation and the enjoyment of this important resource" (MNRaF 2014). The objectives of the Plan include the protection of the unique ecological and historical areas of the Escarpment and the enhancement of the water supply and natural streams, to ensure that outdoor recreation is not hindered; to preserve the natural landscape by instituting compatible farming or forestry operations, making sure that any development complies with the purposes of the Plan; to provide access to the public; and to aid municipalities in the installation of future planning following the NEPDA outline (MNRaF 2014). The accumulation of protective action taken along the Escarpment has resulted in many conservation areas and parks, which has aided in lessening the impact of development; however, even with the continued outpouring of development proposals, the majority of the Escarpment is privately owned land (Preston 2001). The outer boundary distinguished within the Plan is fixed and inflexible and can only be changed by a Plan amendment. The Plan outlines seven land-use designations, and Appendix A illustrates a map of each of these land uses for the section of the Niagara Escarpment that is found in the Niagara Region.

2.3.5 *Greenbelt Act/Plan*

The Greenbelt Act of 2005 enabled the creation of the Greenbelt Plan. This was instituted to protect approximately 1.8 million acres of environmentally sensitive and agricultural land. This land was established within the Golden Horseshoe and the Greenbelt Plan helps to protect this area from urban sprawl and development. The Greenbelt Plan builds on the already existing protection of the Niagara Escarpment Plan and includes many more environmentally significant areas, such as the Oak Ridges Moraine. A council exists for various reasons, such as the administration of the Greenbelt, and also to help with the implementation of the plan. The Greenbelt Plan helps support the policies found within the Niagara Escarpment Plan (The Greenbelt Act 2013).

2.4 Defining Landscape

Landscapes are features of the earth's surface that change continuously, both naturally and anthropogenically. These changes can be thought of as either a deterioration or an improvement; however, what constitutes as deterioration or improvement varies depending on a person's perspective (Antrop 1998). Forman and Godron (1981) define a landscape as "a kilometer[sic]- wide area where a cluster of interacting stands or ecosystems is repeated in similar form." It is stated that a landscape is formed by geomorphological processes and disturbances that function simultaneously within the landscape boundary (Antrop 2014). A landscape disturbance is defined as any disruption from an event, no matter how small, that alters the availability of substrate or the physical environment and its resources that ultimately have an effect on the ecosystem, community, or population (Pickett and White 1985). Disturbances can lead to new landscapes on any scale depending on the amount of disturbance involved. This may also lead to landscape fragmentation. Studies of spatial heterogeneity emerged in the 1980s as landscape ecology became more developed with easier access to data and analysis methods. Spatial heterogeneity simply refers to the abiotic and biotic relationships found in a system and how this will affect its ecology (Turner 2005a). The term landscape ecology was first introduced by Carl Troll, a German biogeographer in 1939, who studied regional geography and vegetation science, but was aided by the newly launched perspective of aerial imaging in the 1950s (Turner 2005b). In Europe, landscape ecology was mainly focused on human activities and used for land-use planning. In North America and Australia, there is a more complex approach that observes the spatial patterns at varying scales depending on the organism (Costanza et al. 2007). Agroecosystems consist of the biologic, economic, and social elements of a landscape. This means that both living and nonliving components of the landscape are important for functionality and spatial study (Hietala-Koivu 2002). In studies of sustainability, focus has largely been on the relationship between complex human and environmental systems (Shaker 2015). Due to the various scales mentioned above, the definition of a landscape is rather difficult to ascertain. The definition depends strongly on the particular study. For example, landscapes can be as small as a homeowner's garden, since the biotic and abiotic relationships can be studied for spatial heterogeneity. In contrast, an entire region can be observed for similar abiotic and biotic components, but at a scale, that makes sense for such a large scope, such as urban sprawl and agricultural lands. It is important to note that for this study, changes in small landscape ecological patterns like those found in a garden would not be observed through satellite imagery and therefore no change would be discerned. This study aims to observe a much larger scope of landscape. This study looks at the overall change of the Niagara Region's urban landscape and whether or not this is affecting the agricultural landscape. Magnitude of change can be observed between the years that are being portrayed as a direct comparison between each decade as well as an overall comparison between 1985 and 2015, which is a 30-year period. Spatial patterns can be defined by two aspects: the first being the spatial units, which are often referred to in the literature as the area, region, patch, or zone, and the second being the boundaries between the different features of the landscapes. This leaves landscapes open to change via many varying avenues. The patch-corridor-matrix model is a useful tool to aid in understanding what comprises a landscape. The matrix is the overall theme of an area of study. For example, land use that is predominantly forested area, with a few burn sites, would be considered a matrix of forested area, with patches of burn area. Corridors are strips of land that are between patches and matrices that contain aspects of the matrix directly adjacent. Corridors can fall into four categories. These include line corridors, strip corridors, stream corridors, and networks. Line corridors include pathways or roadways, perhaps even hedges. Strip corridors are wider than the line corridors and may even include patches of their own that allow for migration between other patches and matrices. Stream corridors run along waterways. Networks are those corridors that intersect and cause a loop somewhere in the matrix (Lausch et al. 2015). Often, line corridors are human-made interceptions among patches. While some corridors allow for easy movement from patch to patch, human-made corridors can pose risks to the species that reside in those patches and rely on movement between patches for nutrients. This can be for animal or plant matter. Corridors have an alternate affect in that they can help eliminate the movement of unwanted species or pollutants toward the natural inhabitants of a patch. For example, streams can prevent nutrient runoff from dirtying the water through a process called siltation, which allows sediments to remain suspended in the water (Devi et al. 2008). With the varying components to landscape mosaics, it is only fair to state that natural movements will constantly occur between patches, corridors, and matrices. These movements can be referred to as flow or flux, and the amount of flow or flux between and within the patch-corridor-matrix model is heavily dependent upon the boundaries found within this model (Kindlmann et al. 2005).

2.5 Data

The United States Geological Survey (USGS) provides an excellent website entitled EarthExplorer. This allows for the user to search for an area and find satellite images that coordinate. When searching for an area, it is important to make sure that the appropriate dates are in place. In the beginning of this study, two images were chosen for each year of 1985, 1995, 2005, and 2015. Originally, one image was from the beginning to the middle of the growing season, and the second image was from the middle to the end of the growing season. This means that all of the images would have been captured between May and October of their respective years. It is also important to ensure that all of the images are from the same time period; otherwise, there would be obvious differences depending on the season that would compromise the legitimacy of the study. This means that eight images were originally requested for download: 1985, 1995, 2005, and 2015. Requesting images can take a little while for approval to download. It depends on what the images are to be used for. Landsat imagery was chosen for this study because it is a North American satellite series developed by NASA, which is world-renowned. Specifically, for the 1985, 1995, and 2005 images, Landsat 5 was used, and for the 2015 images, Landsat 8 was used. Landsat 6 was a failed satellite and Landsat 7 is very commonly used. Landsat 8 is relatively new and thus it was chosen for the final two images for this study. The Landsat series of satellites cover the entirety of the globe, not only the Niagara Region. There are more than 40 years of land-use change that can be seen with Landsat. The large scope applications of these satellites are astounding. The images acquired were the following (Table 2.1).

Table 2.1 Imagery for study area

Datasets – Landsat satellite imagery					
Acquisition date	Satellite & sensor	Spatial resolution	Path	Row	Projection
1985-05-24	5, TM	30 m	017	030	NAD 1983 UTM Zone 17 N
1985-09-20	5, TM	30 m	017	030	NAD 1983 UTM Zone 17 N
1995-05-20	5, TM	30 m	017	030	NAD 1983 UTM Zone 17 N
1995-07-30	5, TM	30 m	017	030	NAD 1983 UTM Zone 17 N
2005-07-02	5, TM	30 m	017	030	NAD 1983 UTM Zone 17 N
2005-09-11	5, TM	30 m	017	030	NAD 1983 UTM Zone 17 N
2015-06-03	8, OLI-TIRS	30 m	017	030	NAD 1983 UTM Zone 17 N
2015-09-16	8, OLI-TIRS	30 m	017	030	NAD 1983 UTM Zone 17 N

The sixth band has a resolution of 120 m because this is the thermal infrared band. It is not useful for the classification of these images as the resolution is not strong enough. The Multispectral Scanner System (MSS) has a much lower spatial resolution of approximately 78 m, and therefore is not used for this study. The OLI-TIRS sensor system used for the final two images are actually instruments that are onboard the Landsat 8 satellite. The Operational Land Imager (OLI) sensor has added two new bands to the satellite. The first is a deep blue band specifically designed for water features. The second is a new infrared band which helps capture cirrus clouds. The TIRS (Thermal Infrared Sensor) adds two new thermal bands to the captured images (United States Geological Survey 2015). To obtain the exact shape of the Region, a boundary shapefile was created using data gathered from the Regional Municipality of Niagara open data website. It was exported into PCI Geomatica and the satellite images were clipped to this boundary once the NDVI (discussed below) equation had been added.

2.6 Methods

The study compares supervised classifications done with satellite imagery throughout the 30-year period with that of land-use data analyzed through the use of landscape metrics. Figure 2.2 outlines how both of these methods were undertaken. Both will be described in great detail to follow.

There are a number of research endeavors that have utilized similar methods to those found in this study; however, none have conducted studies in the Regional Municipality of Niagara for the same 30-year time period. This time period is significant as it directly corresponds with the initialization of the abovementioned Niagara Escarpment protection policies. Studying whether these policies had any effect at a regional level on urban growth is an interesting use of landscape metrics in this Region. The following is a very brief glimpse at some work conducted in recent years that observe many of the techniques used in this paper. This study strives to visualize the difference between the urban and agricultural areas over time. Once this supervised classification was completed, it was apparent that the resolution of the imagery would not be enough to undertake a Category 2 classification scheme for all images. Thus, Category 1 classification schemes were utilized. Category 1 allows for more blatant classification titles, for example, commercial, agriculture, residential, etc. This meant that orchard, vineyard, pasture, etc. would be under the agriculture umbrella, simplifying the classification process. Figure 2.4 shows a Category 1 classification using the June 3, 2015, Landsat 8 imagery. Here, the same aspects of the landscape are highlighted (i.e., residential) without the confusion of multiple classes. In the northern corner of the imagery in Fig. 2.3, it is clear that agriculture is confused between orchard/vineyard and pasture/open land. With the resolution of Landsat 8 imagery, it is very difficult to be accurate that that classification is correct. Figure 2.4 reduces this confusion by enabling an agriculture class as a whole.

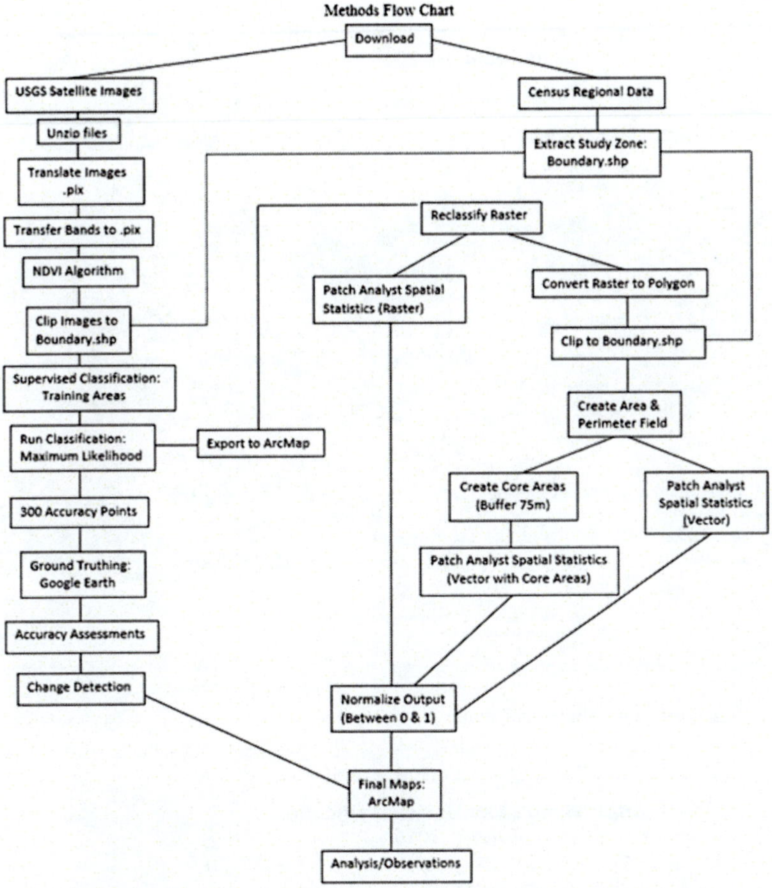

Fig. 2.2 Methodology

Once the classification scheme was clear through trial and error, supervised classifications could be done for all of the images. This was done by creating training sites for each class. The class was selected and training sites could be drawn directly on the image being classified. Once all of the training areas were completed, the classification was run. For the supervised classifications in this study, the maximum likelihood method was chosen, which simply means that if there was a controversy between pixel prioritization, then the pixel would be assigned a class that makes the most sense based on those pixels surrounding it. Once the classifications had been run, a colorized image was presented, such as those shown in Figs. 2.3 and 2.4. Once the classifications were completed, 300 accuracy points were used to ensure the classifications were done as well as possible. When using accuracy points, there is the potential to use another image to match the classes in the image. The author used the original image for reference as well as Google Earth for ground truthing purposes. To ensure that the pixels being observed were being converted to the

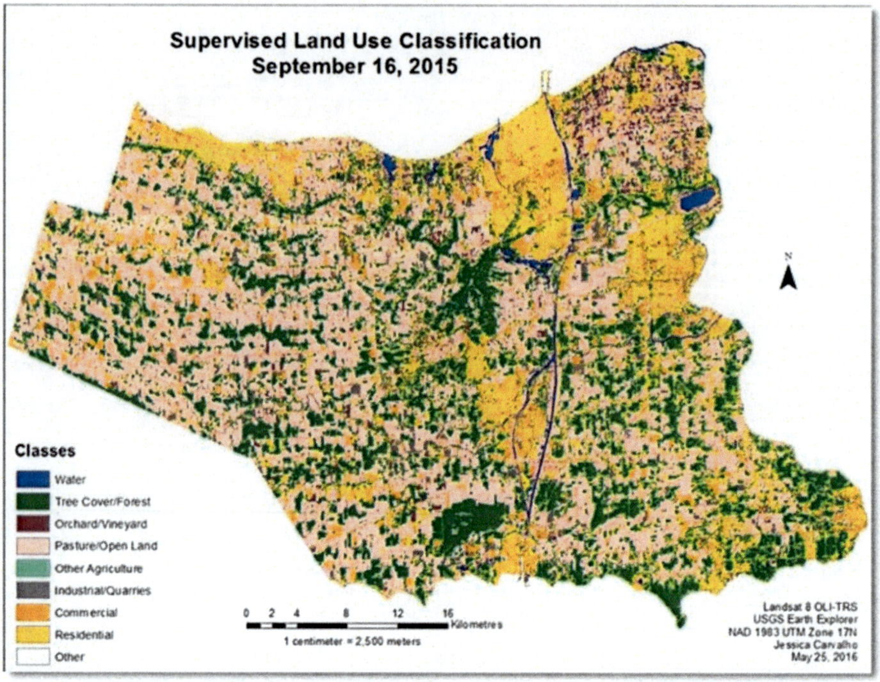

Fig. 2.3 Category 2 classification scheme

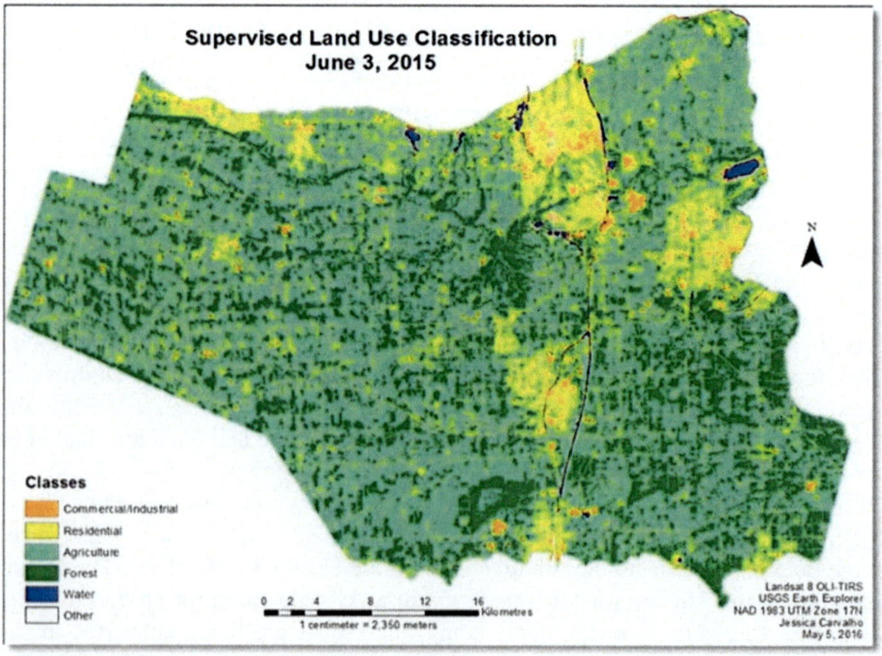

Fig. 2.4 Category 1 classification scheme

appropriate class, ground truthing techniques were applied. Ground truthing for this study can be completed in two different ways. The first is to use a true picture of the area being classified. This will give an indication of which classification each pixel in the area will correspond with. The second is to be present in the location of the study area. Google Earth was the most useful for this study and has been a method utilized by the author in previous studies. Ground truthing using Google Earth is a widely accepted practice throughout the literature. Numerous examples of peer-reviewed articles can be found that have used this method to observe vegetation (Taylor and Lovell 2012; Vega et al. 2011; Duhl et al. 2011). Once the accuracy was completed, it became evident through visual interpretation that there was no need for eight overall image classifications. The images from the same year, but separate ends of the growing season, showed little to no differences. Four images, one from each year of the study, were chosen. The four images were those with the best quality.

2.7 Landscape Metrics

Landscape metrics are a method which includes a series of quantitative indices that are used to describe the patterns of different categorical data (Turner 2005b; McGarigal 2015). This analysis will aid in forming a functional mosaic of land use so that assessment of quantifiable patterns of the landscape can be undertaken. Landscape metrics are useful especially when considering where administrative boundaries lay in comparison to policy changes over time and socio-economic growth or decline (Vaz et al. 2014). Landscape metrics will provide a better understanding of the overall fragmentation of the different land uses, and this study will pay particular attention to the variation of urban growth and rural disturbances (Vaz 2014). Some of the most popular metrics can be seen (Herold et al. 2002). A few examples include the fractal dimension, which measures the perimeter of landscape versus the entire area. Values in this metric range from 1 to 2. If the landscape is less fragmented, it will have a value closer to 1. The larger each perimeter becomes in accordance with the entirety of the area, the closer the value will be to 2. This would indicate a greatly fragmented landscape (Herold et al. 2002). Patch density describes the number of patches within a certain landscape. The size of patches as well as their density can be useful in describing the configuration of the landscape. Patch size is an indicator of species within and on the edge of a given patch and fragmentation among patches can affect those species (Gergel and Turner 2002). In Vaz's (2014) study of Mumbai's mangroves, a number of metrics were used that will also be utilized for this analysis. The reason for this close similarity is due to the nature of the Mumbai mangrove study. In that study, loss and gain of mangroves was the main focus and proponent as mangroves play a large role in biodiversity and overall ecological health of that particular region. In this study, loss and gain of agriculture is the main focus, which also holds a large role in ecological health of the Niagara Region. The loss or gain of agriculture will be ascertained by discerning the growth of the urban landscape. Hence, similar metrics will be utilized as they have been tested in similar literature. It is important to note at which level the metrics are studied. This could mean that metrics are run on a class level, which would include the different classes of the study level (i.e., agriculture, commercial, forest, etc.). This means that metrics would be computed for the whole of the landscape, but by class and not as one overall image.

2.8 Results

While agriculture has grown slightly (0.57% in the study period), there is clearly much more fragmentation. The greater fragmentation observed through the use of these metrics is a cause for concern. Fragmentation allows for isolated parcels of land that maintain habitats, but with greater barriers of urbanized land surrounding them. Agriculture requires large amounts of land in order to operate successfully depending on the agricultural practice. With increased urbanization, farms become increasingly fragmented, meaning that these operations are smaller and become more difficult. This can be caused by roadways or by the outward sprawl of rural towns steadily growing. This has become such an issue recently that the Alberta Agriculture and Forestry government sector recognized this pattern of decreased agriculture and conducted a study to help determine where there was greater fragmentation in their own province. Fragmentation not only affects the land's production capacity but also has an impact on the scenic qualities that the Niagara Region boasts. The parcelization of farmland poses problems such as monitoring crop growth or pest infestations (Brabec and Smith 2002). The Greenbelt Act of 2005 added about 4046.856 ha of provincial protection to Niagara's tender fruit crop area in an effort to curb dangerous urban sprawl habits. Urban growth is necessary in order for the Region to prosper economically. The Greenbelt Act in collaboration with the NEP has put forth stringent regulations in order to ensure that the growth does not impede on the agriculture and rural landscapes. Agriculture is the most common land-use type throughout the Region (Pond 2009b). Expansion into rural areas is inevitable as the Region continues to grow. The goal of the Greenbelt Plan and the Places to Grow Act is to slow the spread of low-density residential housing throughout the Niagara Region as well as the rest of the Greater Golden Horseshoe area. Since low-density residential growth is often the cause of most urban sprawl, this is a necessary step in order to alleviate the pressures of expansion on rural landscapes (Pond 2009a). As of July 2017, the NEP and the Greenbelt Act have both been reviewed and updated to further enhance these regulations and ensure their protective qualities. Arguably, these policies have been effective, as can be seen by the actual small growth that agriculture land has seen. Between 1985, which is shortly after the NEPDA was launched, and 2015, which is 10 years after the Greenbelt Plan was introduced, agriculture has seen a growth of approximately 0.57%. With a projected population increase of 40,800 people between 2011 and 2031, these policies will become crucial. This is a potential 9.3% growth rate in a

20-year period, which is actually a slower growth rate than the rest of Ontario (Niagara Region 2014), however; considering that the growth rate for the 30-year period in this study was very similar (10.99%), the projected population is one that is growing much faster than previously for the Region. This will add the need for greater infrastructure, which must also follow the policies set forth by the Greenbelt Plan and the NEP. Part of the Niagara Region's plan to garner population growth is to introduce greater public transportation, which is also a key factor in the Places to Grow Act (OMoMA 2017). Late into 2016, efforts were made via air travel to reduce the time it took to travel from Toronto to Niagara Falls from 2 h to only 15 min. While this seemed like an exciting leap into better public transportation options for the Region, in reality it was not feasible for the majority of the population. Unless an individual is able to afford the $149 round-trip ticket daily, it is not possible. Not to mention each flight holds only eight people (Smith 2016). Currently, there are GO buses available daily that run from Burlington to Niagara Falls. There is a GO train that can be taken from Toronto to the Burlington GO bus. This service is provided by Metrolinx. By 2023, the train line is scheduled to be extended for daily trips from Toronto to Niagara Falls with stops in between. These are great options, but are the immediate future. It is important to recognize that this Region will continue to grow and need better and improved options. High-speed rail is already in existence throughout the majority of Europe and in China. These have had an effect on time-space shrinkages, especially in China. High-speed rail has created new high-speed rail towns in China that do not follow a sustainable development plan, so if high-speed rail were to be added to Niagara, the sustainable development of urban lands needs to be maintained (Chen et al. 2016). These are all options that could happen in a more immediate time frame. However, in planning for the future, it is necessary to think about what could exist for the Region 30 or more years into the future. The addition of hyperloop technology to the Region could be a possible next step within the next 30–40 years. Hyperloop is a train-like mode of transportation; however, it uses near-vacuum tubes to move pods along a track. Transportation methods are a major consideration when discussing the urban land-use expansion into the Region, as this can have an impact on agricultural lands as well as other natural landscapes.

2.9 Conclusions

The trend toward greater growth will prove to be one of an undying nature. With less expensive housing outside of city centers and easier transit options, urban sprawl is a very likely outcome. Sprawl occurs when there are not rigid growth plans in place for smaller towns that are starting to develop more heavily. Though the Niagara Escarpment has heavy protections in place to ensure the pristine landscape remains intact, these protections do not necessarily cascade into the regional planning of Niagara cities and townships, where the escarpment does not extend. The Greenbelt Plan does have some effect on the non-escarpment areas. Of the 12 towns, townships, and cities found in the Niagara Region, Grimsby and Lincoln have shown the most growth, which is likely due to their closer proximity to large cities such as the City of Hamilton, and only slightly further, the City of Toronto. The use of satellite image classifications in this study proved to be beneficial as land use was easily obtained at a 30 m resolution. A 30 m resolution was adequate in order to garner an overall understanding of land-use change in the Region, which means that a Category 2 classification scheme was not undertaken. A more general sense of the land-use coverage was all that was needed and, therefore, Category 1 was obtained relatively easily. A coarser perspective, although not necessarily entirely representative of the exact surface of the Earth, allows for rapid calculations and understanding of the changes that are happening. The four images that were utilized for further analysis were classified using a supervised classification with an NDVI component. This allowed for the greenness in the image to be more prominent. The finished classifications allowed for change detection practices utilizing image differencing to occur. Here, it was clear that urban change had occurred quite heavily in areas that were directly adjacent to the towns or cities of the Region. The image differencing change confirmation encouraged the use of landscape metrics to further analyze the amount of change that had occurred in the 30-year study period. Landscape metrics are an excellent tool in assessing land cover change and ecology of a specific area. This study can be used as an example for a greater need for strict enforcement of regional planning policies and where they need to be implemented. These metrics show that growth in the urban nature of the region will not be lessening over time and will only increase (Shaker and Ehlinger 2007). This decrease in size, but increase in patches, is indicative of fragmentation and is a crucial motivator for increased policy awareness in the Regional Municipality of Niagara.

References

Antrop M (1998) Landscape change: plan or chaos? Landsc Urban Plan 41:155–161

Antrop M (2014) Interpreting diversity in the European landscape. A comment on perspective essays by Agnoletti and Schnitzler. Landsc Urban Plan 126:81–83

Brabec E, Smith C (2002) Agricultural land fragmentation: the spatial effects of three land protection strategies in the eastern United States. Landsc Urban Plan 58:255–268

Chen H, Sun D, Zhu Z, Zeng J (2016) The impact of high-speed rail on residents' travel behavior and household mobility: a case study of the Beijing-Shanghai line, China. Sustainability 8:1–14

Costanza R, Fisher B, Mulder K, Liu S, Christopher T (2007) Biodiversity and ecosystem services: a multi-scale empirical study of the relationship between species richness and net primary production. Ecol Econ 61:478–491

Devi R, Tesfahune E, Legesse W, Deboch B, Beyene A (2008) Assessment of siltation and nutrient enrichment of Gilgel gibe dam, Southwest Ethiopia. Bioresour Technol 99:975–979

Dickason OP (2010) A concise history of Canada's first nations, 2nd edn. Oxford University Press, Oxford, pp 8–12

Duhl T, Guenther A, Helmig D (2011) Estimating urban vegetation cover fraction using Google earth images. J Land Use Sci 7(3):311–329

European Environmental Agency (2006) Urban sprawl in Europe: the ignored challenge. European Commission, Directorate-General, Joint Research Centre, vol 10, pp 1–58. http://www.eea.europa.eu/publications/eea_report_2006_10

Forman RTT, Godron M (1981) Patches and structural components for a landscape ecology. Bioscience 31(10):733–740

Foster J (2000) Doing less with less: the Ontario provincial government's environmental stream lining has weakened the Niagara escarpment Commission's ability to protect one of Canada's natural treasures. Altern J 26(1):4–5

Furberg D, Ban Y (2012) Satellite monitoring of urban sprawl and assessment of its potential environmental impact in the greater Toronto area between 1985 and 2005. Environ Manag 50:1068–1088

Gergel SE, Turner MG (2002) Learning landscape ecology: a practical guide to concepts and techniques. Springer, New York

Herold M, Scepan J, Clarke KC (2002) The use of remote sensing and landscape metrics to describe structures and changes in urban land uses. Environ Plan 34:1443–1458

Hietala-Koivu R (2002) Landscape and modernizing agriculture: a case study of three areas in Finland in 1954–1998. Agric Ecosyst Environ 91:273–281

Kindlmann P, Aviron S, Burel F (2005) When is landscape matrix important for determining animal fluxes between resource patches? Ecol Complex 2:150–158

Lausch A, Blaschke T, Haase D, Herzog F, Syrbe R, Tischendorf L, Walz U (2015) Understanding and quantifying landscape structure: a review on relevant process characteristics, data models and landscape metrics. Ecol Model 295:31–41

McGarigal K (2015) Fragstats help. University of Massachusetts, Amherst. http://www.umass.edu/landeco/research/fragstats/documents/fragstats.help.4.2.pdf

Ministry of Natural Resources and Forestry (2014) The Niagara Escarpment Plan. https://www.ontario.ca/document/niagaraescarpment-plan

Niagara Escarpment Planning and Development Act (2012). http://www.e-laws.gov.on.ca/html/statutes/english/elaws_statutes_90n02_e.htm

Niagara Region (2011) Niagara Region population 1996–2011. https://www.niagararegion.ca/about-niagara/statistics/population-and-maps.aspx

Niagara Region (2014) Niagara's demographics for housing needs. https://www.niagararegion.ca/social-services/technical-report/demographics-report.aspx

Ontario Ministry of Municipal Affairs (2017) Growth plan for the Greater Golden Horseshoe, 2017. Places to Grow http://placestogrow.ca/index.php?option=com_content&task=view&id=430&Itemid=14

Ontario's Niagara Escarpment (2015a) Canadian biosphere reserves. http://www.escarpment.org/biosphere/index.php

Ontario's Niagara Escarpment (2015b) The Niagara Escarpment Planning and Development Act (NEPDA). http://www.escarpment.org/landplanning/developmentact/index.php

Park Planning Branch (1976) Nature Reserve candidates in the Niagara Escarpment Planning Area. Environ Plan Ser 8(1). Ministry of Natural Resources; Division of Parks

Parks Canada (2005) Creating a sustainable future: a community guide to Canada's Biosphere Reserves. http://www.longpointbiosphere.ca/wp-content/uploads/2013/04/Canadian-Biosphere-Reserves-2005-Your-Biosphere.pdf

Petrov AN, Sugumaran R (2009) Quantifying spatiotemporal dynamics of agricultural landscapes using remotely sensed data and landscape metrics. Geocarto Int 24(3):223–240

Pickett ST, White PS (1985) The ecology of natural disturbance and patch dynamics. Academic, New York, p 472

Pond D (2009a) Institutions, political economy and land-use policy: greenbelt politics in Ontario. Environ Polit 18(2):238–256

Pond D (2009b) Ontario's greenbelt: growth management, farmland protection, and regime change in southern Ontario. Can Public Policy 35(4):414–432

Preston SM (2001) Tracing the connection among municipal amalgamation, community identity, and planning for the Niagara escarpment in Ontario. Environment 29(2):91

Shaker RR (2015) The Well-being of nations: an empirical assessment of sustainable urbanization for Europe. Int J Sustain Dev World Ecol 22(5):375–387

Shaker RR, Ehlinger TJ (2007) Agricultural land fragmentation and biological integrity: the impacts of a rapidly changing landscape on the streams in southeastern Wisconsin. Fish Ecology Laboratory: University of Wisconsin – Milwaukee

Smith M (2016) New flight service connecting Toronto and Niagara to launch in the fall. The Canadian Press

Taylor JR, Lovell ST (2012) Mapping public and private spaces of urban agriculture in Chicago through the analysis of high-resolution aerial images in Google earth. Landsc Urban Plan 108:57–70

The Canadian Commission for UNESCO (n.d.) Biosphere Reserves in Canada. Retrieved 2015. http://unesco.ca/

The Greenbelt Act, 2005 (2013). http://www.mah.gov.on.ca/Page195.aspx

Turner MG (2005a) Landscape ecology: what is the state of the science? Annu Rev Ecol Evol Syst 36:319–344

Turner MG (2005b) Landscape ecology in North America: past, present, and future. Ecology 86(8):1967–1974

UNESCO (2014) MaB: man and the biosphere programme. http://www.unesco.org/new/en/natural-sciences/environment/ecologicalsciences/man-and-biosphere-programme/

United States Geological Survey (2015) Landsat 8 OLI (Operational Land Imager) and TIRS (Thermal Infrared Sensor). Long Term Archive, U.S Department of the Interior. https://lta.cr.usgs.gov/L8

Vaz E (2014) Managing urban coastal areas through landscape metrics: an assessment of Mumbai's mangrove system. Ocean Coast Manag 98:27–37

Vaz E, De Noronha T, Nijkamp P (2014) Exploratory landscape metrics for agricultural sustainability. Agroecol Sustain Food Syst 38:92–108

Vega MB, Craig NM, Lindo GA (2011) Ground truthing of remotely identified fortifications on the central coast of Peru. J Archeol Sci 38:1680–1689

Chapter 3
COVID-19 in Toronto: Investigating the Spatial Impact of Retailers in the Food Retail and Food Service Sector

3.1 Introduction

The novel coronavirus, SARS-CoV-2 (COVID-19), is a major disruptive event that has exposed the global economy to a series of threats and challenges (Hailu, 2020). Initially identified in Wuhan, China, COVID-19, caused by SARS-CoV-2 (severe acute respiratory syndrome), was declared as a global pandemic by the World Health Organization (WHO) on March 11, 2020. It has since spread rapidly around the world, and as of April 27, 2021, over 147.2 million cases have been confirmed. The COVID-19 pandemic has resulted in unprecedented economic crises, disproportionately affecting various sectors of the economy. Certain industries have been hit harder than others, leaving certain retailers more vulnerable to the consequences of the pandemic. Industries that rely on the movement of people have been the most impacted, while those that rely on the movement of information have been left relatively unscathed (Klein and Smith 2021).

The retailing sector makes up an increasingly large portion of the urban economy. The impacts of the COVID-19 outbreak on the food retailing sector in Canada have been significant and, to a great extent, unexpected. In order to slow down the transmission of the virus, governments worldwide have implemented restricted travel and social interactions known as "lockdown measures" as one of the key interventions to contain the spread of the virus. Other imposed regulations include self-isolation, social distancing, curfews, and stay-at-home orders. While these measures are vital to stop the spread of COVID-19, these measures have had a significant negative impact on the economy worldwide (Untaru and Han 2021). Restrictions on movement limited consumer's access to specific goods and services, challenging the stability and testing the resilience of many businesses. The sectors that rely predominately on high levels of human interaction, such as the retailing sector, have experienced the worst impact (Koren and Pető 2020). Due to the consequences of the pandemic and the severity of the health concerns of the COVID-19

E. Vaz, *Regional and Urban Change and Geographical Information Systems and Science*, Advances in Geographic Information Science,
https://doi.org/10.1007/978-3-031-24731-6_3

virus, it resulted in the temporary closure of nonessential businesses. At the start of the pandemic, Canadian grocery stores experienced large-scale panic buying which led to many empty shelves. By the end of March 2020, most stores were closed to walk-in customers with the exception of grocery stores and pharmacies, which implemented strong social distancing rules in their premises. In addition to grocery stores, fast casual restaurants such as Starbucks and other national restaurant chains suspended in-store dining and seating, in favour of take-out, drive-through, and delivery services only.

Despite increasing scholarly interest on COVID-19 and its impact on the performance of food retailers and consumer shopping behavior, there is limited research exploring the spatial risk of COVID-19 on retailers in the food retail and food service sectors. Additionally, much of the current literature has studied the impact of COVID-19 on retail businesses by assessing retail sales data and consumer satisfaction rates during the pandemic; however, impact on location does not exist. As a result, an assessment of COVID-19's spatial impact on businesses based on their location is required to be able to understand which sectors or retailers will need to make strategic adjustments in order to adapt to a rapidly changing marketplace.

In an attempt to fill these gaps in the literature, this study aims to utilize spatial analysis techniques to understand the COVID-19 impact on retail locations. In this proposed research, the impact of COVID-19 on business locations in Toronto will be examined by investigating four retailers from the food retail and food service sectors which are Starbucks, McDonald's, Shoppers Drug Mart, and Loblaws. The aim of this research is to explore the spatial distribution of COVID-19 cases over the past year in Toronto and the changing business dynamics due to the pandemic. The research study is based on a cross-comparative analysis on the distribution of COVID-19 and business density. Starbucks, McDonald's, Shoppers Drug Mart, and Loblaws, which make up a large portion of the economy, will be explored to conduct a cross-economic comparison study on impact. This research seeks to address the following objectives:

1. To determine clusters of positive COVID-19 cases in Toronto and the locations of Starbucks, McDonald's, Shoppers Drug Mart, and Loblaws
2. To determine retail locations that are within a COVID-19 hotspot
3. To explore the socio-economic characteristics that are associated with the identified locations

The hypothesis of this research paper is that based on the number of COVID-19 cases in an area, some retailers will suffer more than others depending on their location and distribution pattern. Understanding the spatial impact on business locations in the food retail and food service sectors in the study area provides important insights into future pandemic-related events, allowing retailers to be prepared for another occurrence because patterns are likely to be similar. This will highlight neighborhoods in Toronto that can be considered at-risk business locations due to their high volumes of COVID-19 cases. Additionally, examining locations of risk against various socio-economic variables will allow retailers to implement strategies to attract customers to continue to shop at their store despite new COVID-19 regulations. Given the contributions of the food retail and food service sectors to the Canadian economy, analyzing the implications and effects of such disruptions on these retailers is critical as it helps understand where the sectors appear to be heading.

The structure of the paper is as follows. The literature review provides a comprehensive review of academics' theoretical and methodological contributions to the theories of the COVID-19 pandemic, changes in the business dynamics and consumer behavior, and its consequences to the food retail and food service sectors. The study area section describes where the study is taking place. The data section is an overview of the data sources used in the study and where it is collected. The methodology section provides a comprehensive description of the steps taken in the study. The results section presents the key results from the kernel density estimation (KDE), Global and Local Moran's I statistic, and geographically weighted regression. The discussion section discusses the findings of the analysis results, answers the hypothesis, explains the limitations, and presents next research steps. The paper concludes with a summary of the research findings.

3.2 Literature Review

3.2.1 Factors Influencing the COVID-19 Pandemic

The coronavirus pandemic has had a significant impact on a global scale. The recent outbreak of COVID-19 has caused many disruptions in the lives of people all over the globe. A number of research studies have assessed the global and regional impacts on epidemiological factors at a spatial scale combining several spatially explicit methods in health geography. Locational information of COVID-19 is critical to understand as it sets a framework for further epidemiological analysis of transmission characteristics (Vaz 2021). Populations of positive COVID-19 cases can be linked to many factors relating to the socio-economic and geodemographic characteristics of a neighborhood. Understanding this information can directly enable mitigation and spatial decision support (Vaz 2021; Shabani and Shahnazi 2020).

According to recent literature, the spatial distribution of COVID-19 is heterogeneous. Penerliev and Petkov (2020), Sannigrahi et al. (2020), Credit (2020), and Vaz (2021) investigate several socio-economic variables that are associated with high cases of COVID-19. Many key variables have been identified as important when understanding the spatial pattern of COVID-19. All of these studies demonstrate population density as a key variable in assessing COVID-19 transmission. According to the studies of Penerliev and Petkov (2020) and Sannigrahi et al. (2020), there is a direct correlation between the age structure of the population and the number of COVID-19 cases and deaths. Infections were more fatal among the population of above-working age. Credit (2020) found that Hispanic and Black neighborhoods had significantly higher case rates compared to neighborhoods that were dominated by non-Black and non-Hispanic racial groups. Furthermore, Vaz (2021) determined that neighborhoods with significant urban footprint and high crime rates were of particular concern for COVID-19 transmission. Transportation, education, income, and social vulnerability were key drivers in his study and were linked to the spread of COVID-19. Similarly, Sannigrahi et al. (2020) concluded that regions in European countries that were dominated by senior and old age population, poverty, and low income were at risk of COVID-19 infections. Through these findings, it is evident that high case rates of COVID-19 are related to factors of environmental injustice, social poverty, and marginalization.

3.2.2 COVID-19 Impact on the Economy and Retail

The COVID-19 pandemic has been one of the biggest disruptive catastrophes in recent decades and has had a global effect on society and the economy. As economic concerns and restrictions continue to affect consumers worldwide, new buying and consumption patterns and behaviors have emerged (Untaru and Han 2021). This rapid shift in social behavior is directly impacting the economy (Kim et al. 2021). The impact on the retail sector is heterogeneous and depends on the combined effect of social distancing and lockdown measures. Retail enterprises are generally dependent on regular consumer traffic and face-to-face interaction, resulting in a severe reduction in market demand caused by the pandemic (Untaru and Han 2021). This is due to the nature of the retail environment. The dense, open, and public characteristics of retail stores were viewed as a major risk for the uncontrollable transmission of COVID-19. As a result, in an effort to minimize the risk of spreading the virus, business closures have become more common. For example, since the pandemic, Starbucks has reinvented its store estate across North America, making the decision in June 2020 to expand convenience-led formats, while closing up to 400 locations (Rushton, et al. 2021).

The impacts of the COVID-19 pandemic on the food retailing and food service sector in Canada continue to evolve with its long-term consequences still unknown. Prior to the pandemic, the food service industry has been rapidly expanding, accounting for more than 30% of the average Canadian's food expenditure (Goddard 2021). Household expenditures on food retail and food service changed significantly in response to public health concerns and government-implemented restrictions (Panzone et al. 2021). The actions taken under the states of emergency changed the reality of food purchasing. Many researchers believed that the shock would result in long-term structural changes, particularly toward online retailing (Rushton et al. 2021). Others predicted that the shock would be largely temporary, suggesting that shoppers would quickly revert to their old habits once restrictions are lifted (Goddard 2020).

3.2.3 Changes in Consumer Shopping Behavior and Consumption

The introduction of social distancing and lockdown measures to contain the spread of the virus imposed strict restrictions on the life of consumers and retailers (Panzone et al. 2021). Recent studies have revealed that the emergence of the COVID-19 pandemic and the extended periods of lockdown and quarantine have fundamentally changed consumer behavior, creating new consumption patterns in response to the perceived threat of the infection (Untaru and Han 2021; Koren and Pető 2020; Brandtner et al. 2021; Panzone et al. 2021).

3.2.3.1 Stockpiling and Hoarding

The shopping behavior resulted in shortages of products in grocery stores (e.g., toilet paper), which exacerbated people's concerns and have increased stockpiling behavior. Brandtner et al.'s (2021) research, which focused on the impact of COVID-19 on customer satisfaction by assessing physical grocery retailing, indicated that product unavailability is among the main factors of customer satisfaction. This uncertainty may lead customers to anticipate future demand by purchasing more than usual resulting in hoarding and stockpiling goods (Panzone et al. 2021). Multiple cases of panic buying and hoarding of nonperishable food items and cleaning and sanitary products have been observed worldwide. Ben Hassen et al. (2020) indicated that American consumers increased their expenditure during COVID-19 in an attempt to stockpile vital household products such as food. Given that food is the most important item, panic buying is a common human response to the crisis that is driven by a fear of running out of food. Furthermore, the act of stockpiling food and other necessities can also be a strategy to reduce the number of future shopping trips (Panzone et al. 2021). According to recent studies (Untaru and Han 2021; Goddard 2020), consumer health consciousness increased as a result of the pandemic as people realized the importance of the immune system and hygiene regarding fighting the virus. Therefore, consumers chose to adopt several health protective measures during shopping activities such as reducing shopping frequencies, minimizing the amount of time spent in stores, avoiding going to multiple stores, standing in crowded places, wearing gloves, and wearing masks in order to stay safe and reduce the risk of contracting the virus (Untaru and Han 2021)

3.2.3.2 Restaurant to Grocery Store Purchases

Furthermore, Goddard (2020) indicated that stay-at-home orders immediately changed the way individuals looked at food purchases. The shift in consumers' food purchasing from restaurants to grocery stores occurred quickly causing disruption in food supply chains. According to Statistics Canada, food expenditures shifted from food service to grocery store purchases on average across the year. Consumers realized their vulnerability to the disease and the loss of employment, causing the change in food habits. With more time spent at home, consumers spent more time cooking, baking, and eating three meals a day at home (Ben Hassen et al. 2020), reducing out-of-home consumption, and experimenting in food purchasing and preparation (Panzone et al. 2021). The shutdown of sit-down dining restaurants had also switched food purchases to grocery stores, significantly changing the volume of types of foods purchased at grocery stores (Goddard 2020). The pandemic in Canada has changed the reality of food purchasing. This caused the food retailing sector to rapidly increase in sales through supermarkets and convenience stores, while the food service sector steadily declines.

3.2.3.3 Transition to Online Purchasing

The pandemic has also introduced Canadians to a multitude of purchasing options. This exposure has generated a demand for "omnichannel" shopping, the ability to shop through different methods such as online, in-store, or pickup, which is likely to be permanent (The Economist 2021). Many brick and mortar retailers adopted new customer servicing options which include curbside pickup and the expansion of home delivery services. Those unable to leverage e-commerce tools have seen their sales decline by up to 89% (Rushton, et al. 2021). Saba (2020), a business reporter from the *Toronto Star*, reported that Loblaws saw its e-commerce business grow 280% during the second quarter of the pandemic, making $1.2 billion in e-commerce sales as of mid-July 2020 compared to $1 billion in all of 2019. As of April 2018, only 28% of 1000 Canadian respondents had shopped for groceries online (Brown 2018). Due to COVID-19 and its impact on consumer behavior, retailers such as Starbucks indicated a desire to power convenience retail with its mobile ordering capability. Starbucks implemented a new strategy to expand pickup stores in dense markets, while the move to curbside, drive-thru, and walk-up windows are focused in suburban areas. Starbucks CEO, Kevin Johnson, said that the strategy "aligns closely with rapidly evolving customer preferences that have accelerated as a result of COVID-19, including higher levels of mobile ordering, more contactless pick-up experiences, and reduced in-store congestion" (Rushton et al. 2021). As a result of the pandemic, online shopping has become increasingly important and is encouraged to limit the spread. The closure of the food service industry has also made online purchasing critical. Therefore, the reality of consumer shopping behavior on food purchases have changed significantly and are heavily impacting retailers.

3.2.4 Applications and Methods Used

3.2.4.1 Qualitative Approaches

Several different methods have been conducted to assess the impact of COVID-19 on retail businesses. Brandtner et al. (2021) and Untaru and Han (2021) conducted an online self-administrated survey questionnaire to determine customer satisfaction due to the pandemic. In addition to collecting data through online surveys, Brandtner et al. (2021) also used external sources like Google review platforms to gather information regarding customer satisfaction levels from selected retail chains in the form of star ratings and textual evaluation comments. This data was explored using appropriate text mining techniques which provided further insights by outlining factors that influence consumer satisfaction. While these studies used data collected through surveys to evaluate the impact of COVID-19 on grocery retailers from the viewpoint of the consumers, Panzone et al. (2021) and Kim et al. (2021) used retail sales data by applying a time series clustering model to determine the impact of changes in sales due to the COVID-19 outbreak. Results showed that some sectors gained, and others lost with significant changes in consumer shopping behavior. In the study of Ben Hassen et al. (2020), it was found that consumers in Qatar preferred online grocery shopping during COVID-19. Panzone et al. (2021) concluded that grocery store chains benefited from the disruption, while restaurants faced significant losses in revenues. Untaru and Han (2021) also revealed that protective measures utilized by retail establishments against COVID-19 exert an influence on customers' satisfaction during shopping.

3.2.4.2 Quantitative Approaches

Spatial techniques assessing the locational impact of the COVID-19 pandemic on business locations are limited; however, there are several different quantitative approaches that have been used to assess the spatial distribution of COVID-19. Rex et al. (2020) conducted an exploratory study, analyzing the spread of positive COVID-19 cases in the State of São Paulo, Brazil, using kernel density estimation (KDE). An extensive number of studies used simple visualization techniques to visualize the spatial distribution of COVID-19, either by choropleth maps or dot density maps. Many researchers utilized spatial autocorrelation techniques such as hotspot analysis, Global Moran's I, and Local Moran's I. Fan et al. (2020) employed hotspot analysis to quantify the spatial distribution of COVID-19 in China. Another research also conducted in China utilizes hotspot analysis to illustrate the spatial intensity of the COVID-19 infection in China (Tang et al. 2020). Similarly, Li et al. utilize Global and Local Moran's I to investigate clusters of COVID-19 incidences in China.

Many research studies have combined geographic visualization of the COVID-19 distribution to additional socio-economic and demographic characteristics. Several different approaches have also been used to explore the demographic and socio-economic variables associated with the spread of the virus. Regression models have been used extensively in many epidemiological studies. Credit (2020) and explored the spatial associations between sociodemographic, economic, and built environmental characteristics with COVID-19 activity using linear regression models. Furthermore, Sannigrahi et al. implemented global and local spatial regression models to assess the spatial association between the driving factors and COVID-19 cases and deaths. Likewise, Vaz (2021) conducted a stepwise backward regression model to establish key variables influencing the spread of the virus. Zhang and

Schwartz (2020) utilize multivariable regression models and conclude that there are strong positive correlations for socio-economic factors including population density, proportions of elderly residents, poverty, and percent population tested with COVID-19 morbidity or mortality. Similarly, Mollalo et al. (2020) applies a geographically weighted regression model to examine COVID-19 incidences in the United States, in relation to socio-economic, demographic, behavioral, topographical, and environmental factors. It was discovered that income inequality played a role in explaining COVID-19 incidence.

3.3 Study Area

The area of study in this paper is the City of Toronto, the most densely populated city in Canada, with a population density of 850 inhabitants per km^2, which is significantly higher than the average for Ontario municipalities. The City of Toronto is a unique study area as it is an urban area with the highest population density in Canada and a high level of socio-economic and cultural diversity. Given its high population density, Toronto is at risk of experiencing an excess of COVID-19 cases, which appears to be borne out of the numbers of infections and deaths in that city. As of April 27, 2021, Toronto has had 139,375 cases of COVID-19 infections and 3030 deaths (Fig. 3.1).

3.4 Data

The data used for this research is gathered from multiple sources and is summarized in Table 3.1. There are three main sets of data used for this analysis: case counts of positive COVID-19 infections, geographic locations of the four retailers (Starbucks, McDonald's, Shoppers Drug Mart, and Loblaws), and the 2016 census data for Toronto consisting of the socio-economic variables of the study area. The data in this study is critical for the purpose of analyzing spatial clusters, distribution of retailers, and relationships among variables.

3.4.1 COVID-19 Data Gathering and Pre-processing

COVID-19 data for Toronto is collected at the neighborhood level from the City of Toronto Open Data Portal (2021). This variable represents the case counts of confirmed COVID-19 infections and deaths, containing information on the date the case was reported, the source of infection (whether it was by travel or close contact), the patient's gender, whether the patient was hospitalized, and whether the case was resolved or not. The cases were downloaded for four different timestamps

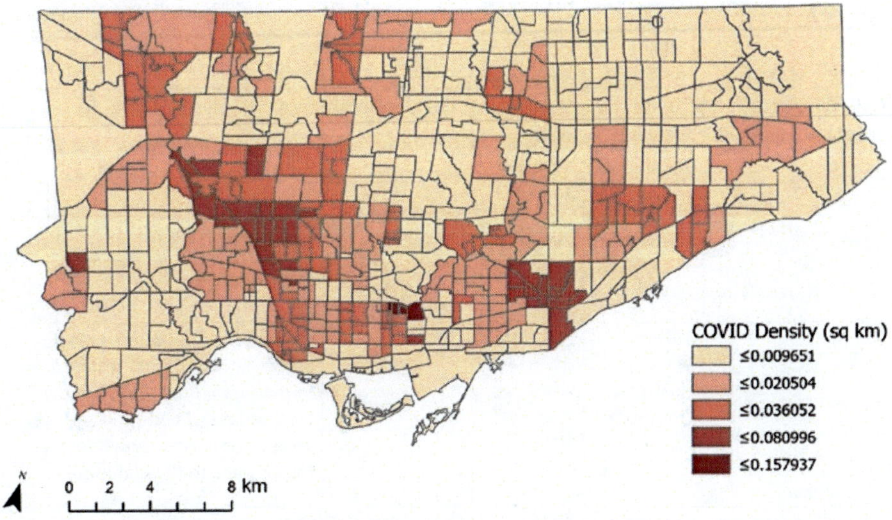

Fig. 3.1 COVID-19 density of cases from May 5, 2021

throughout the pandemic, May 1, 2020, September 5, 2020, January 12, 2021, and May 5, 2021. COVID-19 data is represented as case counts per Toronto neighborhood. To account for the population density, COVID-19 density is calculated using the population density and COVID-19 cases (Vaz 2021). This is represented by the following equation:

$$\mathrm{COV}_d = \frac{\frac{\mathrm{COV}_n}{p}}{A}$$

where COV_d = COVID-19 density, COV_n = number of COVID-19 cases, p = population, and A = area in km^2. Calculating COVID-19 density allows for a more thorough and integrative analysis of the spatial distribution of COVID-19.

Since this data is only made available at the neighborhood level, it is important to spatially downscale this data to census tracts (CTs), so that it can be used to analyze the spatial patterns of COVID-19 across the city at a more detailed spatial unit. This will eliminate the modifiable areal unit problem (MAUP) in the future when other analyses are being conducted at the CT level. It will also help effectively determine the relationships between COVID-19 clusters and retail locations through a spatial comparison. To spatially downscale COVID-19 data, Vaz (2021) conducts a zonal interpolation using population density to generate a higher accuracy of COVID-19 in Toronto. This approach can be implemented in this study to spatially downscale COVID-19 data to the CT level using the Humanitarian Data Exchange (HDX) data for population density in Canada.

Table 3.1 List of variables used

Variable criteria	Data source	Year	Spatial unit	Justification
COVID-19 case counts	City of Toronto	May 1, 2020	Neighborhood level	N/A
		September 5, 2020		
		January 12, 2021		
		May 5, 2021		
Retail data	Yellow Pages Business Directory	2020	CSV	The choice in retailers provides a mix from fast casual restaurants to grocery stores. Being under the food retail and food service sector, they have shown a great importance to the economy
• Starbucks				
• McDonald's				
• Shoppers Drug Mart				
• Loblaws				
Socio-economic variables	CHASS	2016	Census Tract	These variables were chosen because they were proven to have a significant influence on COVID-19 from past research studies. Sannigrahi et al. (2020) and Vaz (2021) found that socio-economic factors such as income, education, and employment were crucial in determining the spatial relation of COVID-19 cases. Regions demonstrating a lower income, education, and employment had higher case rates. Immigration and visible minority population were also identified as important variables in this study
• Low-income status				
• Low education				
• Visible minority				
• Recent immigrants				
• Subsidized housing				
• Unemployment				

The HDX data is collected at a 1km spatial resolution and is extracted for Toronto using the boundary of the city. The population density for Toronto and the Toronto CTs are then used to conduct a zonal statistic, creating a density surface. This is then spatially joined with the case counts for COVID-19 using the field operation First. This results in a local zonal interpolation at the CT level making this substantially better than the standard neighborhood analysis.

3.4.2 Retail Data Gathering

Business locations of the four retailers in Toronto, Starbucks, McDonald's, Shoppers Drug Mart, and Loblaws, are retrieved from Yellow Pages Business Directory and are collected as a CSV file. The reason why retailers from the food retail and food service sector were chosen for this analysis is due to its staple importance in retail. The choice in retailers provides a mix from fast casual restaurants to grocery stores. Being under the food retail and food service sector, they have shown a great importance to the economy. The retail location data retrieved from Yellow Pages include the address, city, province, postal code, and country of each of the Starbucks, McDonald's, Shoppers Drug Mart, and Loblaws locations in Toronto. Data on each of the store locations were then used to geocode using the address locator in ArcGIS Pro into a shapefile of point features containing longitude and latitude coordinates for each retail location.

3.4.3 Socio-economic Data Gathering and Pre-processing

The 2016 census data for Toronto consisting of the socio-economic data of the study area are obtained from CHASS (2016) at the CT level. Census data provides a wide range of information pertaining to the population including social and economic factors.

Once the data is collected, the variables must be rated and standardized. All variables are rated into a percentage. Population variables are rated using the total population and household variables are rated using the total number of households. Rated variables are then standardized using the z-score method to ensure that all variables contribute evenly to a scale. Standardization refers to the process of subtracting the mean and dividing by the standard deviation. Standardized variables are variables that have been rescaled to have a mean of zero and a standard deviation of one. This process can be modeled as:

$$Z = \frac{x - \mu}{\sigma}$$

where Z = standard score, x = observed value, μ = mean of the sample, and σ = standard deviation of the sample. All independent variables are standardized using an automated tool in SPSS.

3.5 Methodology

The methodology for this research study (see Fig. 3.2) is separated into two sections. The first section is an exploratory spatial analysis of the study which includes a visual investigation of the distribution of COVID-19 incidences, comparing four different time periods during the pandemic, and the distribution of the selected retailer locations. With the use of GIS, this study utilizes techniques such as kernel density estimation and further progresses toward testing for spatial autocorrelation using spatial statistics. This exploratory visual investigation may be used to make conclusions about the clusters of confirmed cases of COVID-19 and the locations of Starbucks, McDonald's, Shoppers Drug Mart, and Loblaws. This will help identify

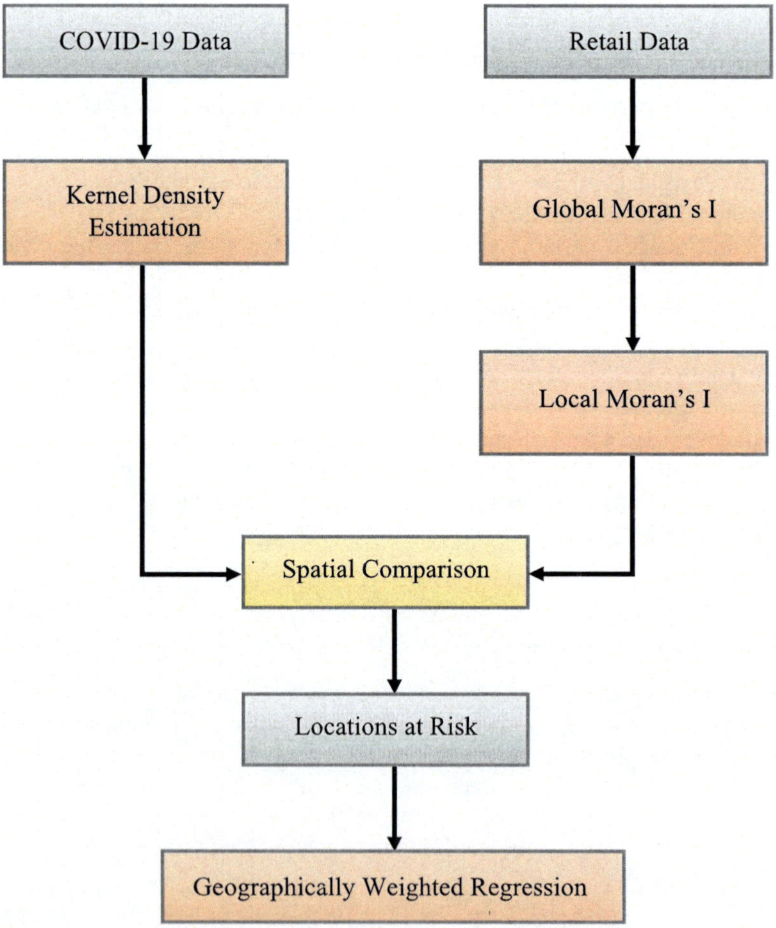

Fig. 3.2 Outline of methodology

retailers that have been impacted the most due to their location based on the distribution of COVID-19 cases across the city.

While the first section of the study provides a general overview of the distribution and overall spatial relationship of COVID-19 and business locations across the city, the second component of this study examines several socio-economic variables that have been proven to have an impact on COVID-19. This will help determine the neighborhood characteristics of the retailers that have and will continue to suffer the most due to the pandemic based on their location.

3.5.1 Kernel Density

To explore and compare the spatial distribution pattern of COVID-19 cases from the four different time periods in Toronto, kernel density estimation (KDE) is applied. The kernel density method is a probability density estimation technique used to identify cluster patterns based on point features. KDE calculates the density of point features around each output raster cell, producing a smoothed curved surface fitted over each point. The following formula models how the kernel density for points is calculated:

$$\text{Density} = \frac{1}{(\text{radius})^2} \sum_{(i=1)}^{n} \left[\frac{3}{\pi} \cdot \text{pop}_i \left(1 - \left(\frac{\text{dist}_i}{\text{radius}} \right)^2 \right)^2 \right]$$

$$\text{For } \text{dist}_i < \text{radius}$$

where $i = 1, \ldots, n$ are the input points, pop_i is the population field value of point i, and dist_i is the distance between point i and the (x, y) location.

This is done in ArcGIS using the Spatial Analyst tool, "Kernel Density," to produce a density estimation of COVID-19 density from May 1, 2020; September 5, 2020; January 12, 2021; and May 5, 2021. Prior to the kernel density estimation, COVID-19 data at the census tract level for each time period is converted into point data using the "Feature to Point" tool in ArcGIS Pro. These points are then used as the input parameter of the Kernel Density function. The population field that is used to weight the features is the COVID density calculations. This weights some features more heavily than others based on its value. The search radius, also known as the bandwidth, is computed for this estimation using the default bandwidth estimation method: Silverman's rule of thumb. This approach of calculating the default radius is resistant to spatial outliers and eliminates the ring around the points phenomena that often occurs with sparse datasets. This process of determining the kernel density estimation is applied for all four time periods and is masked within the Toronto boundary.

3.5.2 Assessing Spatial Autocorrelation: Using Global Moran's I

The Spatial Autocorrelation (Global Moran's I) tool is used to measure the spatial autocorrelation between each of the four selected retailers based on their feature locations and feature values. The Global Moran's I statistic looks at the spatial association of the entire study area where it determines the similarity of one object to another. Given a set of features and an associated attribute, it evaluates whether the pattern expressed is clustered, dispersed, or random. The values of the Global Moran's I statistic range from +1 (perfect cluster) to −1 (perfect dispersion) to 0 (random spatial pattern). The Moran's I statistics for spatial autocorrelation is represented as:

$$I = \frac{n}{S_0} \frac{\sum_{i=1}^{n} \sum_{j=1}^{n} w_{i.j} z_i z_j}{\sum_{i=1}^{n} z_i^2}$$

where z_i represents the standard deviation of an attribute for feature i from its mean $\left(x_i - \overline{X}\right), w_{i,j}$ represents the spatial weight between feature i and j, n represents the total number of features, and S_0 is the aggregate of all the spatial weights:

$$S_0 = \sum_{i=1}^{n} \sum_{j=1}^{n} w_{i,j}$$

Using ArcGIS, the Moran's I index was calculated for the four selected retailers, Starbucks, McDonald's, Shoppers Drug Mart, and Loblaws. Prior to running the Global Moran's I, the number of stores for each selected retailer is aggregated by census tract by spatially joining the point locations of each retailer to the Toronto census tract boundary file. The input parameters for measuring spatial autocorrelation are the census tract shapefile and the count of each retailer. The Moran's I value is also associated with a p-value and z-score to indicate whether or not it is statistically significant in the model. The Global Moran's I index is run using the polygon contiguity edges and corners option for the Conceptualization of Spatial Relationships parameter. This option is used because polygon features are used in this analysis. For Contiguity edges corners, polygons that share an edge or corner will influence computations for the target features.

3.5.3 Local Moran's I Statistic

While the Global Moran's I statistic determines the spatial association of the entire study area, the Cluster and Outlier Analysis also known as the Anselin Local Moran's I (LISA) statistic is applied to visualize the distribution of features that are similar and dissimilar, visually presenting the spatial association with the Moran's I values. Given a set of weighted features and an analysis field, the tool identifies spatial clusters of features with high or low values as well as spatial outliers. It calculates a local Moran's I value, *z*-score, and pseudo *p*-value, representing the statistical significance of the computed index values, and a code representing the cluster type for each statistically significant feature. The local Moran's I statistic is represented as:

$$I_i = \frac{x_i - \underline{x}}{S_i^2} \sum_{j=1, j \neq 1}^{n} w_{i.j(x_j - \underline{X})}$$

where x_i is an attribute for feature i, $\underline{X}$ is the mean of the corresponding attribute, $w_{i.j}$ is the spatial weight between feature i and j, and:

$$S_i^2 = \frac{\sum_{j=1, j \neq 1}^{n} \left(x_j - \underline{X}\right)^2}{n-1}$$

with n equating to the total number of features.

Using ArcGIS, the Cluster and Outlier Analysis tool is applied for each of the retailers (Starbucks, McDonald's, Shoppers Drug Mart, and Loblaws), to identify areas of statistically significant clustering, dispersion, and spatial outliers using the Anselin Local Moran's I statistic. The tool is run using the contiguity edges and corner spatial relationship. A statistically significant positive cluster with a high positive *z*-score indicates that the surrounding features are similar regardless of whether they are clustered with high (high-high) or low (low-low) values. A statistically significant negative cluster with a low negative *z*-score indicates that the surrounding features are dissimilar regardless of whether the neighboring features are high to low (high-low) or low to high (low-high) values.

3.5.4 Geographically Weighted Regression

A geographically weighted regression (GWR) is applied to explore the socio-economic characteristics of the identified COVID-19 hotspots. GWR is a local form of linear regression that is used to model spatially varying relationships. By fitting a regression equation to every feature in the dataset, GWR assesses a local model of the variable being predicted. Separate equations are constructed by incorporating the dependent and exploratory variables of the features falling with the neighborhood of each target feature. The Neighborhood Type and Neighborhood Selection Method parameters determine the shape and extent of each neighborhood analyzed. The dependent variable used for the GWR is COVID density from May 5, 2021, and the exploratory variables are low-income status, visible minority population, recent immigrants, subsidized housing, low education, and unemployment. All variables are rated and standardized. The parameters used for this analysis are Continuous for Model Type, Number of neighbors for Neighborhood Type, and Golden search for Neighborhood Selection Method. When using the Golden Search option, the tool determines the best values for the Number of neighbors parameter using the golden section search method by first determining the maximum and minimum distances and testing the AICc at various distances incrementally between them.

3.6 Analysis Results

This section presents the findings from the spatial and statistical methods applied in this study. It begins by describing the spatial distribution of COVID-19 and business locations in the City of Toronto. It then presents the results of spatial autocorrelation and explores the socio-economic characteristics associated with the COVID-19 hotspots using spatial regression techniques.

3.6.1 Spatial Distribution of COVID-19

Within the City of Toronto, confirmed cases of COVID-19 are unevenly distributed and found primarily in the Downtown core. The following figures resulting from the kernel density estimation of COVID-19 reveals areas where there is a high concentration of cases as well as areas where there is a medium and low concentration of cases. The distribution pattern resulting from the kernel density estimation can be used to analyze the movement of COVID-19 across Toronto throughout the past year starting from May 2020 to May 2021. Throughout the pandemic, it is evident that there are a high number of cases observed largely in Downtown Toronto, and the concentration of cases steadily decreases as you move outside the Downtown core. Looking at all four timestamps, as of May 1, 2020, the highest density of cases is found in Downtown Toronto with a medium-high density of cases found in York and East York, a medium density of cases found along the borders of Etobicoke and North York, and a low density of cases in the east end of Toronto in Scarborough region and some areas in the west in Etobicoke. As of September 5, 2020, after the first wave of the pandemic, medium-high density of cases extends to west and north of Toronto near Etobicoke and North York. There is also a low-medium density of cases present in east end Toronto, extending to Scarborough, along Lake Ontario. After the second wave of the pandemic, January 12, 2021, the density of COVID-19 cases in most sections of Toronto has decreased with a significant concentration of cases remaining in the Downtown core. COVID density has decreased significantly in the west and east ends of Toronto, Etobicoke, and North York. Following the third wave of the pandemic, May 5, 2021, it is clear that the density of cases in the outer areas of Toronto has fallen substantially, with a high density of COVID-19 remaining in the Downtown core.

In all four time periods, there is a high concentration of cases found primarily in the Downtown core, indicating a potential COVID-19 hotspot due to its high population density. The movement of COVID density across Toronto throughout the pandemic poses a potential threat to retailers existing in one of these COVID-19 hotspots. In addition to the Kernel maps indicating areas with the highest occurrence density, it also demonstrates the probability of businesses being negatively impacted by the virus based on their location. Hot areas on the map represent areas where the pandemic will have the greatest impact on retailers, whereas cold areas on the map represents areas where the COVID-19 pandemic will have the least impact on retailers based on their location.

3.6.2 Spatial Distribution of Retailers

The spatial distribution pattern of Starbucks, McDonald's, Shoppers Drug Mart, and Loblaws is explored through global and local indicators. Results from the Global Moran's I aim to describe and quantify the degree of spatial autocorrelation among the store locations of Starbucks, McDonald's, Shoppers Drug Mart, and Loblaws. Table 3.2 summarizes the results from the Global Moran's I statistic. The Moran's I index ranges from -1 (negative spatial autocorrelation) to $+1$ (positive spatial autocorrelation), with values near 0 being closer to a random distribution. Its value is associated with a z-score and p-values and are important to determine the significance of the results.

Referring to Table 3.2, Starbucks exhibits a statistically significant association and a positive z-score value. Therefore, the null hypothesis can be rejected that there is no spatial independence in the distribution of Starbucks but rather a more spatially clustered spatial distribution. This indicates that Starbucks has a distribution that is $<1\%$ likelihood its clustered patterns that are the result of random chance. However, McDonald's, Shoppers Drug Mart, and Loblaws exhibit no statistical significance for its spatial patterning. Its Moran I Index is closer to 1, thus making their distribution of random chance.

Table 3.2 Spatial autocorrelation of retailers

	Moran's index	*z*-score	*p*-value	Pattern
Starbucks	0.327	14.996	<0.001	Clustered
McDonald's	0.018	0.836	0.403	Random
Shoppers Drug Mart	0.036	1.591	0.112	Random
Loblaws	−0.019	−0.753	0.452	Random

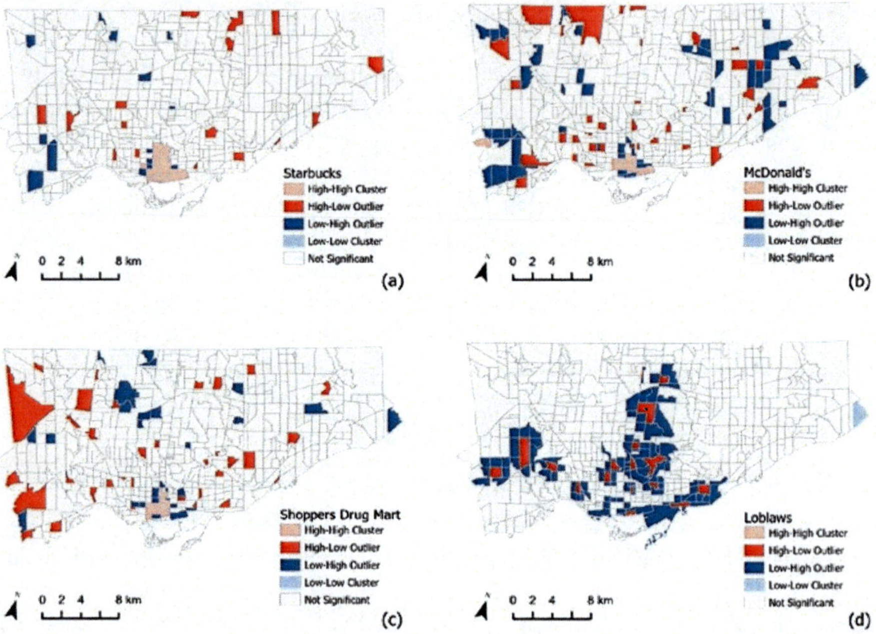

Fig. 3.3 (**a**) Hotspots and coldspots of Starbucks. (**b**) Hotspots and coldspots of McDonald's. (**c**) Hotspots and coldspots of Shoppers Drug Mart. (**d**) Hotspots and coldspots of Loblaws

Results from Local Moran's I further visualize the distribution of significant and insignificant spatial autocorrelation for each retailer. Figure 3.3a–d display the results from the Cluster and Outlier Analysis tool, displaying hotspots and coldspots of each retailer. Census tracts in shades of red represent positive spatial autocorrelation of high-high clusters or high-low outliers, while census tracts in shades of blue represent negative spatial autocorrelation of low-low clusters or low-high outliers. Starbucks, McDonald's, and Shoppers Drug Mart display a statistically significant cluster of high values (high-high cluster) in the Downtown core, indicating that the surrounding features have similar high values. Starbucks has the largest cluster present in Downtown Toronto which is explained by the Global Moran's I index that indicated a clustered pattern. These retailers do not have any low-low clusters present in the dataset. While retailers are mainly clustered in the downtown region, spatial outliers of high-low and low-high values are present throughout the entire study area indicating that there is an outlier in which a high value is surrounded primarily by low values (high-low) and an outlier in which a low value is surrounded primarily by high values (low-high). Contrarily, Loblaws does not exhibit any high-high clusters. As indicated from the Global Moran's I results, the distribution of Loblaws stores demonstrate a random pattern. Due to its random pattern, all Loblaws locations either are represented as a spatial outlier of high-low values or low-high values or are not significant on the map.

3.6.3 *Cross-Comparative Analysis of COVID-19 and Retailer Distribution*

Starbucks

Within the City of Toronto, locations of Starbucks coffee are highly populated in Downtown Toronto where there is also a high density of COVID-19 cases (see Fig. 3.4). These census tracts have the highest number of Starbucks of 11, 12, and 16 locations present in each CT. Locations of Starbucks are less clustered moving away from the Downtown core; however, these census tracts do have at least more than one Starbucks located in them. These areas also have a medium-high COVID

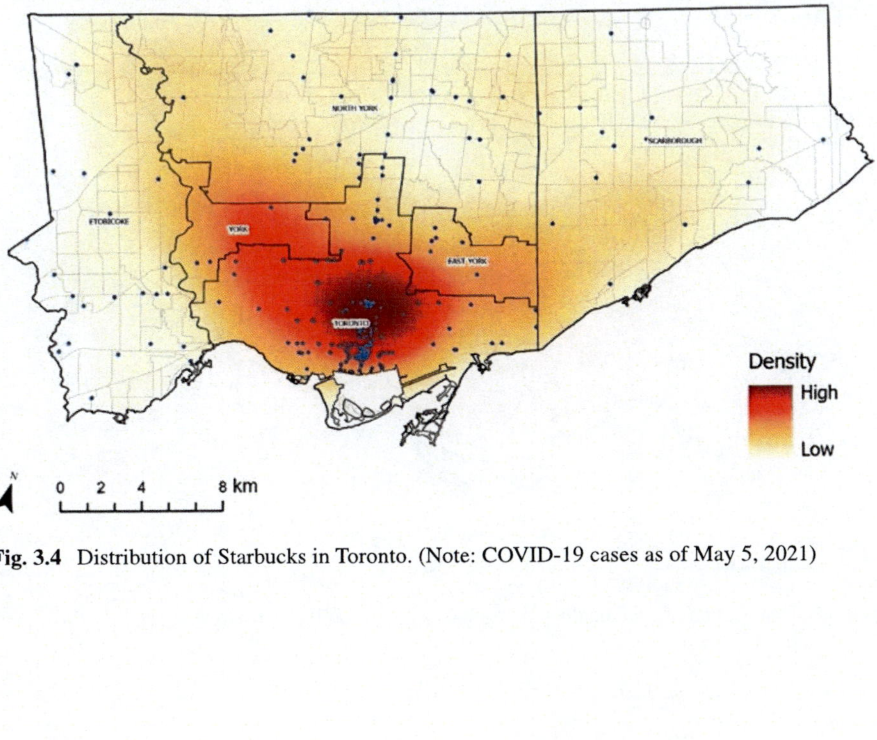

Fig. 3.4 Distribution of Starbucks in Toronto. (Note: COVID-19 cases as of May 5, 2021)

density. In the peripheral areas of Toronto, Starbucks locations are scattered around in the outer edges of Etobicoke, North York, and Scarborough where there is a low COVID density and are clustered along the TTC subway line where there is a low to medium density of COVID-19 cases.

The downtown core can be identified as high risk due to its high concentration of COVID-19 cases and Starbucks locations. There are over 50 Starbucks locations present in this hotspot and they are of high risk. Census tracts in York and east of Downtown Toronto are medium-risk locations for Starbucks as they have a medium-high density of COVID-19 cases and a medium concentration of Starbucks locations. Starbucks located in census tracts in North York are at low-medium risk due to their low-medium density of COVID-19. These areas also have some Starbucks located in them and could be of potential risk due to the lower density hotspots found in the outer areas of North York. In the outer edges of Toronto (Etobicoke and Scarborough), there is a lower impact for Starbucks as there is a low density of COVID-19 incidences.

McDonald's

Similar to the Starbucks locations, a portion of the McDonald's locations are concentrated in the downtown core of Toronto where there is a high density of COVID-19 cases (see Fig. 3.5). Although Downtown Toronto has a cluster of McDonald's locations, it is significantly fewer than Starbucks with an average of two McDonald's locations in each census tract in the downtown core. McDonald's locations in a COVID-19 hotspot in the Downtown core are at high risk due to their location. The McDonald's that are located east of Toronto, in York, and along the edge of Scarborough are at medium-high and low-medium risk because there is a high-medium concentration of cases in these areas. Although these McDonald's are

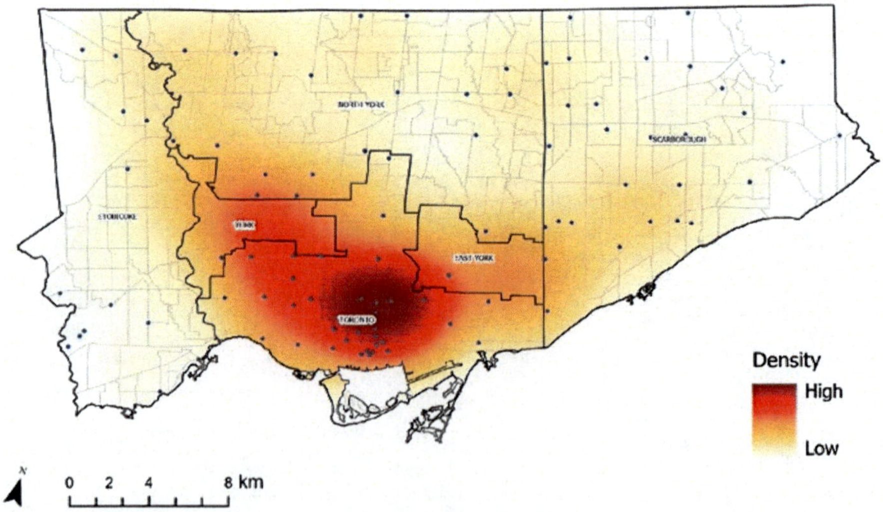

Fig. 3.5 Distribution of McDonald's in Toronto. (Note: COVID-19 cases as of May 5, 2021)

located in areas that have a medium concentration of COVID-19 cases, in the peripheral areas of Toronto, locations of McDonald's are scattered around the city. Due to their random pattern, in regions outside of the Downtown core, these locations are less likely to be impacted because they are not clustered in one area. There are fewer McDonald's locations that are present in a COVID-19 hotspot than Starbucks locations. Furthermore, results from the Global Moran's I statistic confirms that locations of McDonald's are distributed at random even though there is a small cluster of stores present in Downtown Toronto. Locations that exist on the outer edge of Etobicoke and Scarborough are at the lowest risk due to their low density of COVID-19.

Shoppers Drug Mart

Similar to McDonald's, locations of Shoppers Drug Mart are concentrated in the Downtown core and are scattered in the peripheral areas of the city (see Fig. 3.6). The areas in which Shoppers Drug Mart is the most concentrated are also areas with a high density of COVID-19. Census tracts belonging in the COVID-19 hotspot in Downtown Toronto contain over 25 Shoppers Drug Mart stores and are the most impacted by the COVID-19 pandemic. In the peripheral areas of the city, Shoppers Drug Mart locations are scattered across the city with no distinct pattern. This is also confirmed by the Global Moran's I statistic that the spatial distribution of Shoppers Drug Mart locations exhibits a random pattern even though it shows a small cluster of stores in the Downtown region. Similar to the other retailers, Shoppers Drug Mart located in the outer areas of Etobicoke and Scarborough are the least impacted by COVID-19 due to its low density of cases in these areas. Shoppers Drug Marts that are located in York and the shore of Scarborough are at

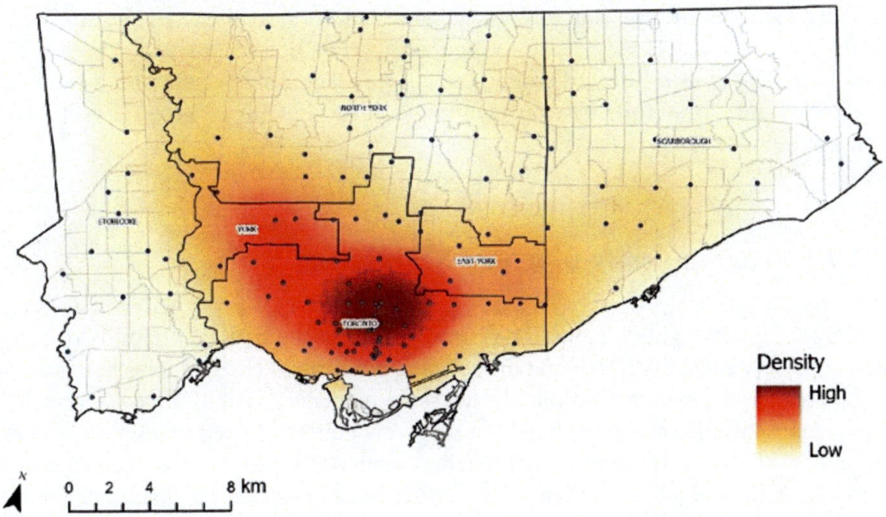

Fig. 3.6 Distribution of Shoppers Drug Mart in Toronto. (Note: COVID-19 cases as of May 5, 2021)

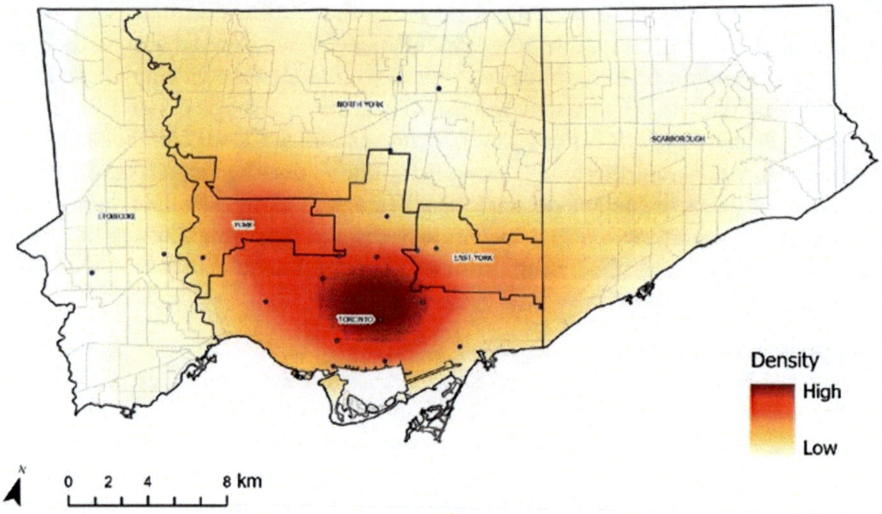

Fig. 3.7 Distribution of Loblaws in Toronto. (Note: COVID-19 cases as of May 5, 2021)

medium-high risk due to its medium concentration of cases. Shoppers Drug Mart locations in North York are at medium risk of being exposed to COVID-19 because census tracts in these areas have a medium density of COVID-19 cases.

Loblaws

Contrary to Starbucks, McDonald's, and Shoppers Drug Mart, locations of Loblaws are not clustered in Downtown Toronto (see Fig. 3.7). There is only one Loblaws store that is directly located in the downtown region which is also a hotspot for COVID-19. This means that this location of Loblaws is impacted the most by COVID-19. The Loblaws that are located around this hotspot are at medium-high risk since they are still located around the hotspot. The Loblaws that are located in Etobicoke and North York are impacted the least by COVID-19 because of its low concentration of cases in these areas.

3.6.4 Geographically Weighted Regression

Geographically weighted regression is used to explore the spatial variability of relationships between COVID-19 and the selected predictors. The variables being used to predict the dependent variable are low-income status, visible minority population, recent immigrants, subsidized housing, low education, and unemployment at the CT level. The regression model resulted in an AICc value of 4842.2. This is a measure of the model performance. The AICc value produced for this model is the smallest among all other models, indicating that this model provides a better fit to the observed data. The R-squared is a measure of goodness of fit. The GWR model results in a R-squared value of 0.76. This indicates that approximately 76% of the total variability of COVID-19 density in Toronto can be explained by the predictor variables, low-income status, visible minority population, recent immigrants, subsidized housing, low education, and unemployment.

Figure 3.8 illustrates the derived standardized residual values obtained from the GWR analysis. The residual map is an indicator of the model performance. It provides additional spatial and nonspatial information about the variation in the dependent variable that cannot be explained by the independent variables. There are seven classes in total with the green indicating the maximum residual values by standard deviation and purple indicating the minimum residual values by standard deviation. This map indicates the census tracts in which the regression model overpredicted or underpredicted COVID density. The areas in green indicate underpredictions where the actual values for COVID density are higher than the model estimated. The areas in purple indicate overpredictions where the actual values for COVID density are lower than the model estimated. The areas in white indicate the locations where the actual values and the values estimated from the model have little to no difference. There is a lot of spatial variability in the outer edges of Toronto in Etobicoke, Scarborough, and North York, while there is almost little to no variability or almost no difference at all in the central regions of Toronto. Areas in Toronto that contained a high concentration of COVID-19 were explained by the predicted variables. These are census tracts in Downtown Toronto, York, and East York where there is a strong correlation between COVID-19 and low-income status, visible minority population, recent immigrants, subsidized housing, low education, and unemployment. Although there are some areas in these regions where COVID-19 was either underpredicted or overpredicted, it is however not as significant as areas in Etobicoke,

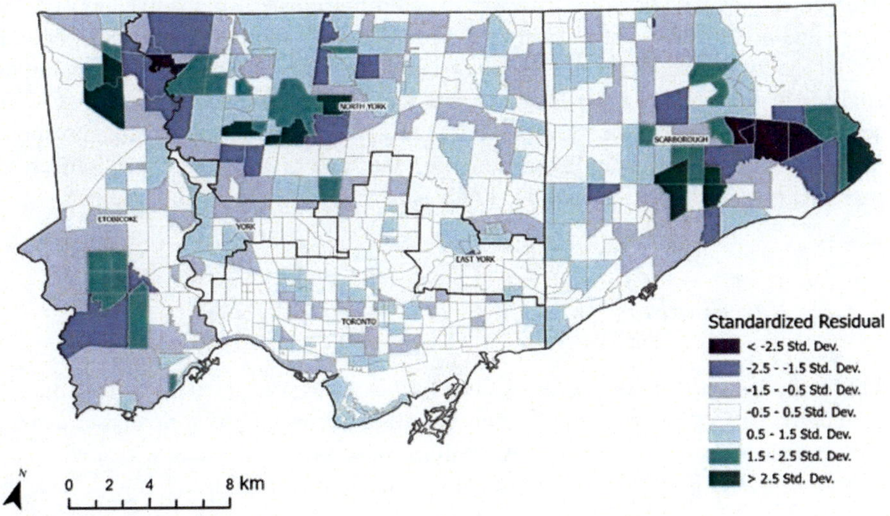

Fig. 3.8 Geographically weighted regression of COVID-19

North York, and Scarborough. These are also areas containing a medium or low concentration of COVID-19 cases.

3.7 Discussion

Understanding the relationship between the spatial pattern of the COVID-19 pandemic and the distribution of Starbucks, McDonald's, Shoppers Drug Mart, and Loblaws in the City of Toronto is increasingly important as the dynamics of the COVID structure continue to change as the pandemic evolves. Although there are many existing literatures on COVID-19 and its impact on the food retail and food service sectors, very few investigate the impact on location and the characteristics of the locations at risk. Therefore, this paper aimed to address these gaps in the literature through spatial analysis techniques to understand the impact of COVID-19 on Starbuck's, McDonald's, Shoppers Drug Mart, and Loblaws locations. The spatial relationships between COVID-19 and business locations in the food retail and food service sectors provide important insights on patterns and connections present in the city as well as into future pandemic-related events. This allows retailers to be prepared for another occurrence because patterns are likely be similar. Additionally, assessing the spatial risk of coffee chains and grocery stores against various socio-economic characteristics can highlight inequalities that are present within a region and strategies to attract consumers to continue to shop at their location despite the pandemic.

The purpose of this study is to use geospatial analysis techniques to assess the spatial distribution of COVID-19 and its impact on Starbuck's, McDonald's, Shoppers Drug Mart, and Loblaws locations. This study looked to investigate three main objectives. The first objective was to determine clusters of positive COVID-19 cases and analyze the distribution of the four retailers through kernel density and global and local indicators. The second objective was to conduct a cross-comparative analysis to determine retailer locations that were impacted the most by the COVID structure by comparing the movement of the COVID-19 pandemic and the hotspots of Starbuck's, McDonald's, Shoppers Drug Mart, and Loblaws. The third objective was to explore the socio-economic variables of the identified locations at risk.

3.7.1 Key Findings

This study has shown that positive COVID-19 cases are clustered in the Downtown core of the City of Toronto, and the concentration of cases steadily decreases as you move outside the Downtown core. A medium concentration of cases is located in the northern parts of Etobicoke and North York and extends from the downtown core to

Scarborough, along Lake Ontario. The Global and Local Moran's I statistics of the spatial distribution of Starbuck's, McDonald's, Shoppers Drug Mart, and Loblaws locations reveal that Starbucks experiences the most impact from the COVID-19 pandemic in the Downtown core. This is due to its clustered pattern resulting from the Global and Local indicators. Majority of the Starbucks locations across the City of Toronto are in a COVID-19 hotspot with over 50% of its locations located in Downtown Toronto. Although distribution of McDonald's and Shoppers Drug Mart stores demonstrated a random pattern as indicated by the Global Moran's I statistic, almost 30% of McDonald's and Shoppers Drug Mart locations are found in Downtown Toronto. These locations also experience the highest impact as they are also located in a positive cluster of COVID-19 incidences. Loblaws, having the lowest number of locations across Toronto, experiences the same amount of impact on its stores located in the Downtown core. However, the spatial structure of Loblaws is quite random and does not indicate any form of clustering, which this means that this retail chain is the least affected by the COVID-19 pandemic as it has only less than 10% of its stores located in a known hotspot for COVID-19.

The analysis was successful in revealing the locations with the highest concentration of COVID density and the retailers that are impacted the most. It is evident that the affected areas have accrued more impact on retailers instead of areas with less COVID cases. As a result, Starbucks experiences the most impact based on its location in the city, while Loblaws experiences the least impact. It is also important to note that fast casual restaurants (Starbucks and McDonald's) experience a significant decrease in sales as consumers start to buy less coffee during the pandemic. Past research suggested that the pandemic changed the reality of food purchasing. Restaurant purchases have been gradually decreasing. With more time spent at home, consumers spent more time cooking (Ben Hassen et al. 2020), thus reducing out-of-home consumption (Panzone et al. 2021). Additionally, results from the geographically weighted regression of COVID density across the City of Toronto suggest that locations at risk are associated with disadvantaged areas with low socio-economic status. Retailers located in disadvantaged areas are more likely to be impacted by COVID-19 as compared to those in more affluent areas. Particularly, areas with low socio-economic status and having a low education, low income, and low employment rate are at higher risk. However, it is evident that the Downtown core has a diverse socio-economic status as shown by the GWR results. Although there is diversity in economic status in the Downtown region, highly developed, high-density areas are also at-risk locations for businesses as transmission rates are more likely to be high when people are near each other. Therefore, continuing to implement other modes of shopping such as curbside pickup, drive-thru, mobile order, and home delivery services (such as Uber eats or skip the dishes) that require less contact and easy access to their food can allow consumers to continue to purchase from their store.

3.7.2 Contributions

By using spatial analysis techniques, this study produces new knowledge on the locational impact of COVID-19 on retailers in the food retail and food service sectors across the City of Toronto. The use of spatial analysis to assess the spatial risk of COVID-19 on business locations and the socio-economic status of these locations facilitate the monitoring of governmental risk measures to contain the disease and to evaluate their efficiency. This study demonstrates the need to prioritize retailers in COVID-19 hotspots to implement strategies to attract consumers to continue to purchase from their location despite new pandemic-related regulations.

In response to the declaration of public emergency status and the changes in consumer behavior, many actions have been taken by the food service, food retail, and the government to ensure the safety of their consumers. As grocery store demand started rising, grocery store chains and supermarkets hired more employees and also provided additional support to employees by raising wages (Goddard 2020). Grocery store chains have also installed protective plastic shields to protect cashiers and adopted customer management strategies to maintain safe distances between customers. This includes limiting the number of customers in a store, encouraging customers to make payments by cards, and marking distances at which consumers waiting in line should adhere to (Untaru and Han 2021). Food service restaurants have also adopted contactless delivery or pickup options. These extra protective measures taken by grocers and food service have also led to decline in their revenue. Despite COVID-19 lockdowns, reported that Canada's largest grocers saw its revenue rise during the second quarter; however, profits declined due to high pandemic-related costs including employee pay raises. Loblaw Companies Ltd.'s revenue increased in the second quarter of the pandemic to 7.4% compared to last year which is almost $12 billion. However, the company that owns the Loblaws supermarket chain and Shoppers Drug Mart saw its net earnings drop 29% over the same quarter last year to $266 million. As reported in the *Toronto Star*, this was due to pandemic-related costs, which included a one-time incentive of $25 million for retail and distribution center workers, as well as $180 million spent on a temporary pay raise for staff. The corporation spent a total of $282 million on safety measures, including the temporary wage increase (Saba 2020).

3.7.3 Limitations

One of the limitations of this study was data availability. Exact case counts of COVID-19 are not available for Toronto at the CT level. If this information were available, a more in-depth examination of COVID-19 density could be used to strengthen the study's findings. This will also eliminate the limitations of the modifiable areal unit problem (MAUP) because all other data used in this study was collected at the CT level. Therefore, to minimize the amount of distortion in this study, zonal interpolation was utilized to generate a higher accuracy of data by combining Humanitarian Data Exchange (HDX) data for population density and COVID-19 data at the neighborhood level.

Another limitation was the presence of ecological fallacy. The assumption is made that all retailers existing in a COVID-19 hotspot will experience the same impact. This applies to the inferences that were made based on the spatial comparison of the kernel density estimation of COVID-19 cases and the retail locations. It is not ideal to conclude that all store locations of a retailer in a COVID hotspot are subject to experience the same impact. This is because there are so many other factors that need to be considered, like the retailers' target market. For example, it is not certain that all Starbucks locations present in a COVID hotspot will be affected the same. Some locations will decline, while some will continue to thrive. This will depend on their exact location in the hotspot. If a Starbucks is located near public transit or in a corporate building, they are likely to be visited more than others because of the commuters and corporate workers in the city.

3.7.4 Future Research and Next Steps

While exploratory, this study examines the social and economic determinants of health in locations of risk due to the current pandemic. To further expand on this research, environmental determinants such as air pollution, crime, and violence could be explored as these have been proven to be indicators of at-risk communities due to marginalization from past research. Further investigation with these variables can help draw clear conclusions on the assumed spatial risk of Starbuck's, McDonald's, Shoppers Drug Mart, and Loblaws.

3.8 Conclusion

This preliminary investigation of COVID-19 using kernel density estimation suggests that the spatial distribution of COVID-19 incidences in the City of Toronto is uneven, with a high density of cases present in the Downtown core. Through Global and Local indicators, this study also revealed that the spatial pattern of Starbucks, McDonald's, and Shoppers Drug Mart is spatially clustered in areas with a high concentration of positive COVID-19 cases. Results determine the spatial risk of COVID-19 on business locations, revealing that Starbucks experienced and will continue to experience the most impact from the pandemic due to its spatially clustered infrastructure in the Downtown core. McDonalds and Shoppers Drug Mart experience the second most impact, while Loblaws experiences the least impact due to its random distribution. Results from the geographically weighted regression model indicate a high correlation with COVID-19 and low socio-economic status further allowing for a better understanding of the characteristics associated with the retail locations at risk from COVID-19. Areas of low socio-economic status with low income, low educations, and low employment rates experience the highest number of cases and have the most impact on retailers.

The findings of the present research indicated the critical importance of the spatial risk of COVID-19 and its locational impact on retailers in the food retail and food service sectors across the City of Toronto. While this research study has given an insight into Toronto's food industry, the findings in this research could also be applied within markets beyond Toronto. Through the assessment of Starbucks, McDonald's, Shoppers Drug Mart, and Loblaws, it is vital to note that COVID-19 does not only have an impact on retailers across Toronto but rather a global impact on retailers worldwide. Beyond the current crisis's public health challenges, governments, international agencies, and businesses have all struggled in different ways to respond to this global risk.

As one of the primary interventions to halt the spread of the virus, governments around the world have put restrictions on travel and social activities known as "lockdown measures." Self-isolation, social distancing, curfews, and stay-at-home orders are among the other imposed regulations. While these precautions are necessary to prevent the spread of COVID-19, they have had a substantial negative impact on the global economy and businesses worldwide. As economic concerns and restrictions continue to affect consumers worldwide, new buying and consumption patterns and behaviors have emerged (Untaru and Han 2021). Recent literature revealed that social distancing and lockdown measures have fundamentally changed consumer shopping behavior and consumption. It has been estimated that worldwide e-commerce increased by 20% in 2020 (Avşar 2021). Anticipating the changing consumer preferences and demands has been a major challenge for marketers.

Epidemics have a disproportionate effect on cities. The current pandemic has been met by unequal responses in different countries (Skegg et al. 2021), leading to unequal impacts worldwide. From this present research, it is clear that retailers clustered in COVID-19 hotspots are hit the hardest. As a result, it is possible that in a post-pandemic spatial equilibrium, the agglomeration premium diminishes, and retailers find it less attractive to locate in cities (Koren and Pető 2020). This pandemic has allowed businesses to evolve and learn from the pandemic, providing important insights into future pandemic-related events, allowing retailers to be prepared for another occurrence. Although COVID-19 vaccines are being distributed in a number of countries, this does not mean the pandemic has been resolved. Cases of COVID-19 continue to rise and lockdown measures in some countries are still in place. Such conclusions drawn from this research can provide retailers from different markets around the world to prepare for the longer-term implications and remediate the shorter-term shocks. It is known that COVID-19 has caused the retail industry to change immensely. However, the long-term social, economic, and health impacts of the virus are still unknown. Retailers that are unsuccessful in adapting to the changes caused by the pandemic will suffer. Through this research, it is easy to understand which sectors or retailers need to make strategic adjustments in order to adapt to a rapidly changing marketplace. Retailers will need to implement strategies to attract consumers to continue to shop at their store despite the risk of COVID-19 and its new regulations. It is critical for businesses to keep an eye on potentially high-impact events in the short term as well as the longer-term landscape in order to establish and maintain resilience. This is important for businesses and the global community to navigate the risks and opportunities ahead in a sustainable manner, strengthen their resilience to future shocks, and progress toward long-term prosperity.

References

Avşar BZ (2021) Omni-channel trends in COVID-19 pandemic. J Int Trade Logist Law 7(1):97

Ben Hassen T, El Bilali H, Allahyari MS (2020) Impact of COVID-19 on food behavior and consumption in Qatar. Sustainability 12(17):6973

Brandtner P, Darbanian F, Falatouri T, Udokwu C (2021) Impact of COVID-19 on the customer end of retail supply chains: a big data analysis of consumer satisfaction. Sustainability (Basel, Switzerland) 13(1464):1464. https://doi.org/10.3390/su13031464

Brown D (2018) Amazon leads online grocery shopping in Canada: survey. Canadian Grocer, April 23. http://www.canadiangrocer.com/top-stories/amazon-leads-online-grocery-shopping-in-canada-survey-79916

Coronavirus (2021) World Health Organization. Retrieved April 27, 2021, from https://www.who.int/health-topics/coronavirus#tab=tab_1

Credit K (2020) Neighbourhood inequity: exploring the factors underlying racial and ethnic disparities in COVID-19 testing and infection rates using ZIP code data in Chicago and New York. Reg Sci Policy Pract 12(6):1249–1271. https://doi.org/10.1111/rsp3.12321

CTV News (2021) Tracking every case of COVID-19 in Canada. CTV News, April 27. https://www.ctvnews.ca/health/coronavirus/tracking-every-case-of-covid-19-in-canada-1.4852102

Fan J, Liu X, Pan W, Douglas MW, Bao S (2020) Epidemiology of coronavirus disease in Gansu Province, China, 2020. Emerg Infect Dis 26(6):1257

Goddard E (2020) The impact of COVID-19 on food retail and food service in Canada: preliminary assessment. Can J Agric Econ:1–5. https://doi.org/10.1111/cjag.12243

Goddard E (2021) The impact of COVID-19 on food retail and food service in Canada: a second assessment. Can J Agric Econ:1–9. https://doi.org/10.1111/cjag.12282

Hailu G (2020) Economic thoughts on COVID-19 for Canadian food processors. Can J Agric Econ 68:163–169. https://doi.org/10.1111/cjag.12241

Kim D, Yun J, Kim K, Lee S (2021) A comparative study of the robustness and resilience of retail areas in Seoul, Korea before and after the COVID-19 outbreak, using big data. Sustainability 13:3302. https://doi.org/10.3390/su13063302

Klein A, Smith E (2021) Explaining the economic impact of COVID-19: core industries and the Hispanic workforce. 1–18

Koren M, Pető R (2020) Business disruptions from social distancing. PLoS One 15(9). https://doi.org/10.1371/journal.pone.0239113

Mollalo A, Rivera KM, Vahedi B (2020) Artificial neural network modeling of novel coronavirus (COVID-19) incidence rates across the continental United States. Int J Environ Res Public Health 17(12):4204

Panzone LA, Larcom S, She P-W (2021) Estimating the impact of the first COVID-19 lockdown on UK food retailers and the restaurant sector. Glob Food Secur 28(Complete). https://doi.org/10.1016/j.gfs.2021.100495

Penerliev M, Petkov V (2020) Geodemographic aspects of Covid-19. Espaço e Economia. Revista brasileira de geografia econômica:18

Rex FE, Borges CADS, Käfer PS (2020) Spatial analysis of the COVID-19 distribution pattern in São Paulo State, Brazil. Ciência & Saúde Coletiva 25:3377–3384

Rushton D, Sillitoe B, Schober S, Woolley J, Bertini P (2021) A playbook for post-pandemic retail site selection: using advanced analytics to drive brick and mortar retail planning in 2021 & beyond. CARTO: report

Saba R (2020) Loblaw profit falls on 'hero pay' and other pandemic-related costs despite surging revenues. Toronto Star, July 23. https://www.thestar.com/business/2020/07/23/loblaw-profit-falls-on-hero-pay-and-other-pandemic-related-costs-despite-surging-revenues.html

Sannigrahi S, Pilla F, Basu B, Basu AS, Molter A (2020) Examining the association between socio-demographic composition and COVID-19 fatalities in the European region using spatial regression approach. Sustain Cities Soc 62(Complete). https://doi.org/10.1016/j.scs.2020.102418

Shabani Z, Shahnazi R (2020) Spatial distribution dynamics and prediction of COVID-19 in Asian countries: spatial Markov chain approach. Reg Sci Policy Pract 12(6):1005–1025. https://doi.org/10.1111/rsp3.12372

Skegg D, Gluckman P, Boulton G, Hackmann H, Karim SSA, Piot P, Woopen C (2021) Future scenarios for the COVID-19 pandemic. Lancet 397(10276):777–778

Tang W, Liao H, Marley G, Wang Z, Cheng W, Wu D, Yu R (2020) The changing patterns of coronavirus disease 2019 (COVID-19) in China: a tempogeographic analysis of the severe acute respiratory syndrome coronavirus 2 epidemic. Clin Infect Dis 71(15):818–824

The Economist (2021) The importance of omnichannel strategies. The Economist: special report, March 13. https://www.economist.com/special-report/2021/03/13/the-importance-of-omnichannel-strategies

Untaru E-N, Han H (2021) Protective measures against COVID-19 and the business strategies of the retail enterprises: differences in gender, age, education, and income among shoppers. J Retail Consum Serv 60(Complete). https://doi.org/10.1016/j.jretconser.2021.102446

Vaz E (2021) COVID-19 in Toronto: a spatial exploratory analysis. Sustainability 13(2):498. https://doi.org/10.3390/su13020498

Zhang CH, Schwartz GG (2020) Spatial disparities in coronavirus incidence and mortality in the United States: an ecological analysis as of May 2020. J Rural Health 36(3):433–445

Chapter 4
Spatial Analysis of Characteristics and Influencing Factors of Killed or Seriously Injured Persons from Motor Vehicle Collisions Within the City of Toronto

4.1 Introduction

As populations continue to grow, the demand for reliable and safe infrastructure becomes an increasing consideration (Aldegheishem et al. 2018). Planners, policy makers, and governments take on these challenges on a regular basis taking into account the balance between safe and efficient transportation (Ewing et al. 2003). The outcome of these decisions is influenced by stakeholders evaluating the current and future demands of a particular road network and comparing that to the demand of its surrounding populations (Witzell 2019). This approach is effective in new or redeveloping areas but becomes a more complex issue in areas where infrastructure change is not easily obtainable. These areas include established urban environments with dense grid networks of roadways that exist and may be characterized by high vertical densities which afford physical barriers for change.

In 2018, a total of 110,114 collisions occurred in Canada resulting in 162,341 personal injury victims and 1922 fatalities (Transport Canada 2018). The number of fatalities observed in 2018 increased by 3.6% in comparison to 2017 with a total of 1856 recorded. According to reports provided by Transport Canada, persons 65 years or older accounted for the highest portion of fatalities representing 22% (430) of all fatal injuries in 2018. Moreover, in all injury classes (fatal, serious, or minor injury) persons aged 25–34 represented the highest or second highest cohort susceptible to injury accounting for 16–19% of total injuries. The World Health Organization (2005) reported that the majority of motor vehicle collisions occur within vulnerable road users such as "pedestrians, cyclists, and motorcyclists." Transport Canada's Canadian Motor Vehicle Traffic Collision Report (2018) identified fatalities by specific road users, observations indicated 67.6% of fatalities involved a driver or passenger (49.9% and 17.7%, respectively), and 30% were related to vulnerable road users (pedestrians, 17.3%; motorcyclists, 10.4%; and cyclists, 2.3%).

E. Vaz, *Regional and Urban Change and Geographical Information Systems and Science*, Advances in Geographic Information Science,
https://doi.org/10.1007/978-3-031-24731-6_4

While physical design may play a critical role in evaluating the level of safety for its users, there is potential for underlying factors to exist and may influence the likelihood for injury along a given road network. This research aims to explore these so-called underlying factors by investigating the spatial-temporal patterns of motor vehicle collisions (MVCs) and evaluate the spatial variation within collision locations. This assessment will measure the level of spatial autocorrelation present to determine whether spatial interaction occurs randomly in nature or represents an underlying pattern that is statistically significant.

Interaction can vary across space, because as units interact with each other, their associated relationships can differ. Describes this dynamic as a "spatial process" which is affected by underlying spatial properties as a unit experiences a change in state. This spatial process is comprised of four types, two of which are most relevant in this research: dispersion and interaction. As a spatial unit experiences dispersion, the underlying spatial structure is impacted. This impact can be helpful in determining the level of randomness that may exist within a given collision cluster. In other words, if the locations of MVCs are observed to have no spatial autocorrelation, the spatial dynamics are deemed to be random.

4.1.1 Study Area

The City of Toronto is municipality located within the southern region of Ontario and is also the province's capital. Toronto has a total population of 2,731,571 and represents over 20% of the entire population of Ontario. Between 2011 and 2016, Toronto has experienced a total population growth of 4.5% or ~1% per year. Estimates provided by Statistics Canada expect populations to rise to 3.73 million by 2046, an increase of 36.6%. It is important to note that Toronto is the largest city in terms of population within Canada, followed by Montreal, QC, and Calgary, AB. The city spans 43 km east to west and 21 km north to south representing a total area of 641 Sq Km (City of Toronto 2020). It ranks among the top ten highest population densities in Canada with an average of 4334 people per square kilometer. There are a multitude of cities that neighbor Toronto including Mississauga and Brampton to the west, Vaughan and Markham to the north, and Pickering to the east.

4.2 Literature Review

4.2.1 Socio-economic Factors

In the United States, fatal traffic collisions have been observed to be a leading cause of accidental death and are most likely to fall within the top ten attributing factors of accidental death across North America.[1] There is a variety of research and

[1] https://wwwnc.cdc.gov/travel/yellowbook/2020/travel-by-air-land-sea/road-and-traffic-safety

literature available in relation to vulnerable road users as these groups are most likely to sustain serious or fatal injury when involved in a collision (Scheidt et al. 1995). These findings increase the reliance for proactive decision-making on law enforcement, infrastructure development, and traffic safety programs (Nunn and Newby 2014). Safety programs can be effective in changing human behavior to reduce the risk of injury and promote a safe and efficient environment for motor vehicle transportation (Pratte 1998).

Research by Camilloni, Farchi, and others (2013) indicated that socio-economic status (SES) played a significant role in influencing the spatial characteristics of where fatal or serious injuries can occur as a result of MVCs. The authors observed individuals with lower income were more susceptible to injury resulting from a MVC and likely explained by the group's inability to access safer transportation. Other research identified a linkage between certain occupations within a lower socio-economic status to be correlated with an increase in injury or fatality likelihood in relation to MVCs (Cubbin et al. 2000). This relationship of groups within lower SES can possibly be explained by their lack of capital freedom causing them to live further from urban inner cities, requiring greater distances to be traveled by car. Alternatively, for groups within lower SES that do live within the inner-city environment, they may be characterized by not having the ability to access safe and independent transportation methods. This lack of access causes a reliance on public transportation and increases the susceptibility to injury as they will likely require more frequent travel as a more vulnerable road user (pedestrian, cyclist, etc.).

4.2.2 Spatial Autocorrelation Techniques

In order to ascertain if a particular unit experiences a spatial phenomenon that is deemed not to be random, statistical assessments are required (Waller 2005). These assessments can be in the form of spatial autocorrelation measurements that compare the spatial interaction of a unit against its neighbors to understand if the observation is statistically significantly different (Anselin 1995). When these results yield a statistically significant finding, conclusions can be made that autocorrelation may exist and the observation is interacting in a manner not deemed to random in comparison to its neighbors. Assessments such as Getis-Ord Gi* statistic (Ord and Getis 1995) or kernel density (Flahaut et al. 2003) are some of the most commonly applied spatial evaluations to identify hotspots.

While kernel density is a widely used approach, Getis-Ord is a preferred method. This is driven by the different ways in which it conceptualizes results; kernel density relies on raw point distributions, while Getis-Ord Gi* statistic has the ability to evaluate the aggregation of points by a particular aggregate unit, therefore allowing associations to be made with additional datasets. Though the two approaches have been proven to have similar results under certain conditions, the objectives of this research require statistical comparisons with additional data repositories supporting the selection of Getis-Ord. Outcomes of this analysis produce two important metrics for final assessments, z-score and confidence level, and the combination of the two allows for an understanding of the spatial interaction and the statistical significance of the observation. More importantly, these assessments reveal insights that are unable to be identified from global spatial autocorrelation measures.

The proposed research aims to explore the spatial patterns of killed or seriously injured (KSI) to understand their relationship to its surroundings and measure the associated spatial interaction that exists within the system. This approach has been leveraged by a variety of researchers and spatial statisticians utilizing hotspots to identify patterns (Flahaut et al. 2003). As a result, outcomes can provide insights on certain geographic areas that could benefit from improved road safety and reduced injury propensity. A study conducted by Martin et al. (2000) evaluated spatial interaction on a temporal scale to support future decision-making on improving road safety and provided an avenue for optimized injury prevention.

4.3 Methodology

It is imperative all measurements and assessments are evaluated in an equal and unbiased manner. The approaches outlined in this section ensure inferences observed are the most reliable in supporting the acceptance or rejection of the null hypothesis. For the purposes of this research, the null hypothesis states:

H_o: The nature of collisions are random and their locations are not influenced by surrounding factors such as environmental attributes, underlying demographics, and the density of infrastructure.

The evaluation of this hypothesis will be tested by leveraging a few different approaches: some spatial, temporal, and statistical.

Statistical bias can occur when observations are unequally compared or when expected values are skewed to misrepresent their true underlying parameters. The methods outlined below will ensure statistical bias is accounted for by using an equal aggregation unit. Hexagons will be leveraged and sized at 500 m in diameter to best identify potential clusters at a scale relevant to road segments and intersections. Comparatively, this technique will provide an unbiased analytical approach while avoiding arbitrary spatial units of measure such a dissemination areas (DA), census tracts (CTs), or other administrative boundaries. These arbitrary systems tend to follow geographic features such as roads, water bodies, and/or land features causing a variety of bias concerns in the size and location of such aggregate units. These challenges can lead to inadequate findings and poor inferences at the lower scales required for this research. Evaluations at this level will also aid the identification of relevant road segments and offer the ability to make connections with independent factors. These factors will facilitate a relationship that aims to explain the physical, environmental, or influential characteristics represented within a given observation.

Moreover, all additional data sources (outlined in Sect. 4.3.1) will be summarized at the same hexagonal units to provide a consistent and repeatable environment across all observations within the study area. Neutrality of analysis will also be obtained following the hexagonal method and improves spatial insights in relation to specific road segments, corridors, and/or regions. Lastly, an increase in measurable spatial interaction can be obtained between spatial neighbors when using hexagons. Its improved spatial complexity offers a 50% increase (six edges vs. four edges) in potential interaction, following a rook contiguity, in comparison to standard square grid systems.

The methods outlined in this section will provide an overview of the various techniques and approaches utilized in this research. Data will be evaluated and processed from five unique sources (outlined in Sect. 4.3.1). Each repository will have a different aggregation technique based on the type of information being summarized, and these range from raw counts, (proportional) coverage analysis, and length analysis (see Fig. 4.6). All data will then be summarized to a uniform hexagonal grid system (outlined in Sect. 4.3.2) to allow for comparisons and spatial autocorrelation testing utilizing both Global Moran's I and Local Getis-Ord Hotspot Analysis. Significant clusters will be evaluated for each region within the City of Toronto and discussed further in Sects. 4.4.2 and 4.4.3. After significant clusters have been identified which exhibit the most concern for injury propensity, a stepwise linear model will be used to extract the strongest predictors likely to influence the rate of KSI (outlined in Sect. 4.4.4) (Fig. 4.1).

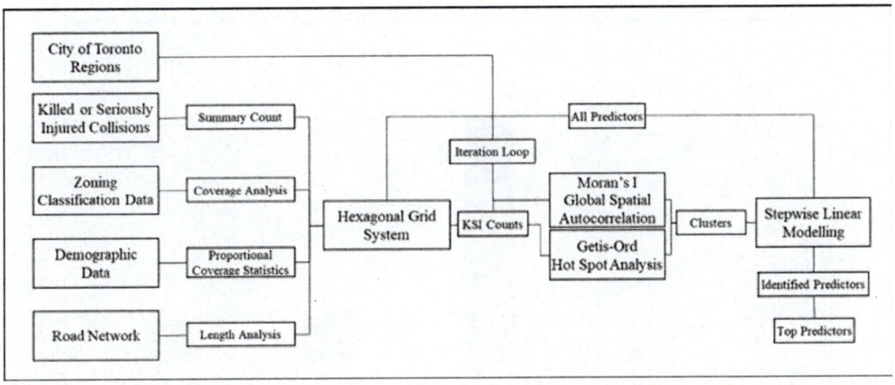

Fig. 4.1 Workflow and methods overview

4.3.1 Data

There are a variety of explanatory factors obtained within this research to allow for testing and understanding of their relevance to injury susceptibility/detection. Factors were sourced through a variety of repositories; each will be expanded in the following section and includes:

(i) Killed and seriously injured (KSI)
(ii) Zoning by-law classifications
(iii) Demographic variables
(iv) Roadways
(v) Former Toronto municipal boundaries

4.3.1.1 Killed and Seriously Injured (KSI)

The KSI data is sourced from the Public Safety Data Portal (PSDP) published by Toronto Police Service (TPS). This dataset includes all killed or seriously injured collisions reported to TPS from 2006 to 2019 and will be the foundation for which all analysis and observations rely on. Between 2006 and 2019, a total of 8416 persons were involved in a MVCs that resulted in fatality or serious injury. Major injuries were observed to represent the highest portion of all KSIs at 64% (5375) followed by minor injuries representing 15% (1264), and minimal injuries at 12% (996). In addition, there were 781 cases (9%) in which a fatality was a result of a motor vehicle collision (see Fig. 4.2). In the lens of Vision Zero, this is 8416 instances where an injury could have been avoided through the adoption of increased

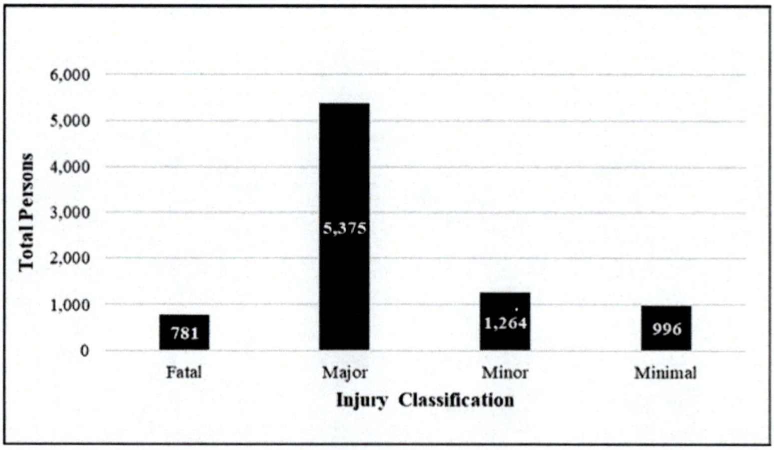

Fig. 4.2 Total KSI persons from 2006 to 2019 by injury severity. (Source: Toronto Police Service, PSDP)

awareness, public safety campaigns, infrastructure improvements, and enforcement tactics to decrease negative driving behaviors (Evenson et al. 2018).

4.3.1.2 Zoning By-Law Classifications

Zoning classifications were sourced from the City of Toronto Open Data Portal. This repository provides a detailed identification of all types of land use within a given region. This product will permit a valuable examination of authorized zoning types defined under Chapter 569–2013 from the City of Toronto Zoning By-Law. Zoning by-laws regulate the use, size, height, density, and location of buildings within the limits of Toronto (City of Toronto 2020). This examination will be vital in supporting the test of the null hypothesis to ascertain if a particular zoning classification has an impact on the density or clustering of KSI occurrences. There are 27 different categories assigned under this dataset, with the top five classifications representing 62% of all the Toronto area (see Fig. 4.3).

Of the top five classifications, two are related to residential use, one open land use, one commercial use, and one not classified. This dynamic is important to mention as these types of uses will likely have the highest influencing factor on injury rates since they account for the highest variation within the dataset.

4.3.1.3 Demographic Variables

Demographic variables will be extracted from the 2016 Canadian Census Profile provided by the Canadian Socio-Economic Information Management System (CANSIM) by Statistics Canada. CANSIM is comprised of over 70 million time series records within the areas of economic, social, financial, and monetary themes (Government of Canada 2021). This data repository will be retrieved using the Computing in the Humanities and Social Sciences (CHASS) Data Centre hosted by the University of Toronto. This data will be of significant value in measuring the

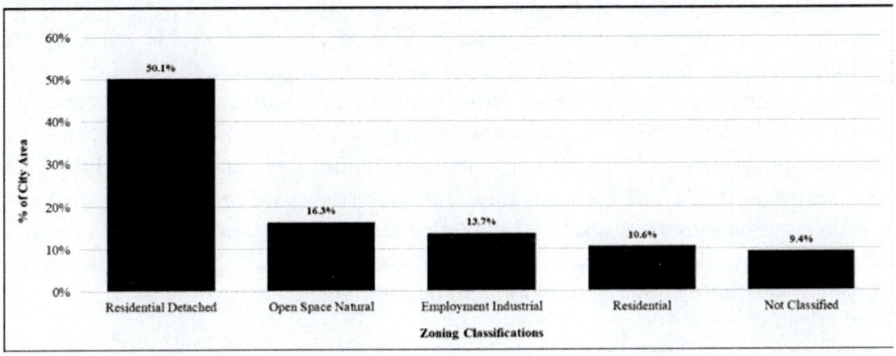

Fig. 4.3 City of Toronto zoning classification (top five). (Source: City of Toronto Open Data Portal)

importance of sociodemographic variables in influencing the density or clustering of KSI occurrences.

This research will evaluate over 1200 variables within the following areas to ensure a highly efficient and comprehensive approach. These include the following topics:

(i)	Population and dwellings size	(vii)	Language
(ii)	Age and sex	(viii)	Income
(iii)	Dwelling characteristics	(ix)	Immigration status/origin
(iv)	Family characteristics	(x)	Education
(v)	Marital status	(xi)	Labor force
(vi)	Household characteristics	(xii)	Journey to work

4.3.1.4 Roadways

Since motor vehicles utilize infrastructure and roadways as a means to move conveyances efficiently from one point to another, it is important that the spatial connections and network densities are accounted for when modeling KSI occurrences. Road networks will be sourced directly from City of Toronto's Open Data Portal and will represent all geographic linear features such as:

(i)	Access road	(xi)	Major arterial
(ii)	Busways	(xii)	Major railway
(iii)	Collector	(xiii)	Major shoreline
(iv)	Creek/tributary	(xiv)	Minor arterial
(v)	Expressway	(xv)	Minor railway
(vi)	Ferry route	(xvi)	Other
(vii)	Geostatistical line	(xvii)	Pending
(viii)	Hydro line	(xviii)	River
(ix)	Laneway	(xix)	Trail
(x)	Local	(xx)	Walkway

See Appendix *for a full list of definitions*

These linear features will be leveraged to assess their potential predictability in determining the density and clustering of KSI occurrences across the City of Toronto. In observing Fig. 4.4, it is apparent that local roads make up the highest portion of infrastructure within the city limits. Local roadways can be defined as neighborhood streets with max speed limit ranging from 15 to 40. Local roadways alone accounted for 39% of all linear features within this dataset and exhibited a total length of 3320 KM for over 25,000 unique streets (see Fig. 4.4). Four additional classifications were observed to be the most representative of the total length which include collector, trails, major arterial, and expressway classifications, cumulatively representing 33% of the total length within the dataset.

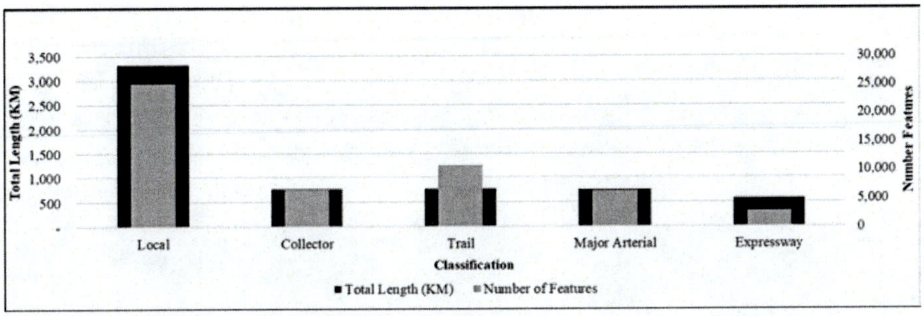

Fig. 4.4 Linear feature classifications for Toronto by total length and feature count (top five). (Source: City of Toronto Open Data Portal)

4.3.1.5 Former Municipality Boundaries

Prior to the agglomeration of what is now known as the City of Toronto, it was previously comprised of six former municipalities: Etobicoke, North York, York, Toronto, East York, and Scarborough. While these regions no longer represent individual municipal governments, they are still referred and referenced by Torontonians (City of Toronto 2020). Considering these municipalities are easily distinguishable, it will allow for improved ability to evaluate how KSI observations change between regions and enables comparison of their temporal characteristics. For the purposes of this research, York, East York, and North York will be consolidated into one region known as "York"; this will ensure comparable distributions exists among regions. As visualized in Fig. 4.5, York represents the largest municipality in terms of total area (232 Sq Km, 35% of the entire region) and is the second highest in total KSI occurrences (2441, 29% of all KSIs). Comparatively, the central municipality, Toronto, represented the highest KSI occurrences with 2651 (32% of all KSIs) occurrences but ranked lowest in total area (111 Sq Km, 17% of the entire region) of the four regions. These early findings infer 61% of KSIs occur within 52% of the entire region, indicating a unique set of characteristics within these zones that differ from the rest of the system.

4.3.2 Aggregation Techniques

Considering supplementary data sources abide by dissimilar aggregation units, proportional coverage analysis will be utilized to ensure all attributes are joined to the same hexagonal unit. This technique also aims to address the problem when visualizing or analyzing data by administrative areal units such as CTs or DAs. When using conventional methods, such as choropleth mapping on unequal spatial units, it is not possible to accurately depict the distribution within a single areal unit without the use of normalization. Proportional coverage analysis provides a solution for this, as it is a type of areal interpolation where data from one set of geographic zones

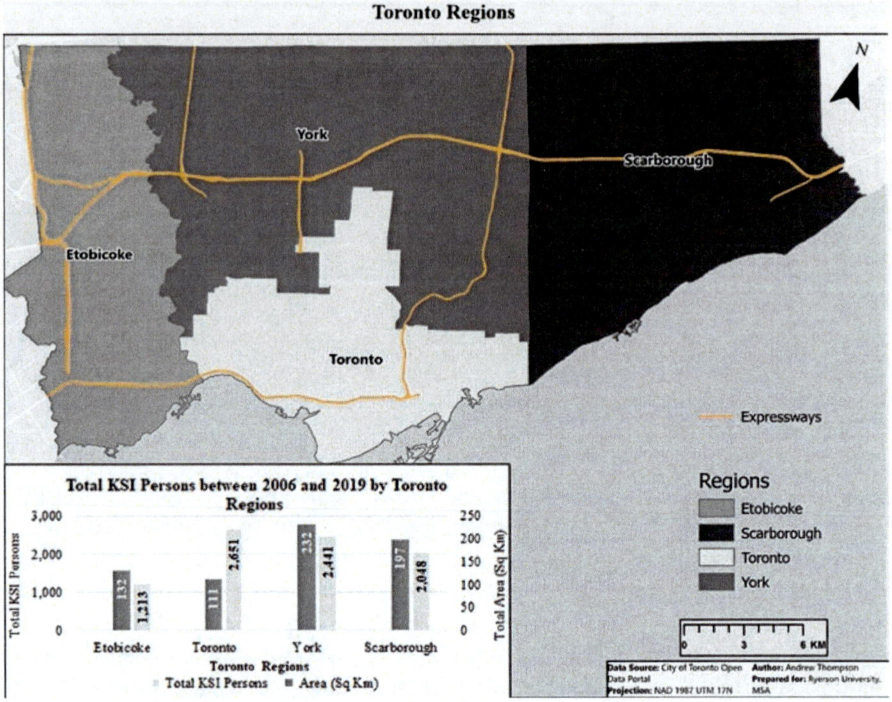

Fig. 4.5 Map of Toronto regions and KSI rates

is transferred to a set of target zones based on their distinct geographic units. The two layers needed for execution include a population layer, such as occurrences of KSI, and an ancillary layer, such as demographics, land-use zoning, and infrastructure type/density. This will be the foundation to the extraction and interpolation method for additional data sources derived from varying aggregate units.

As depicted in Fig. 4.6, the population of a given hexagon can be calculated using the following equation, $p_i(a_i/x_i)$ where p is the total area of the population unit (dissemination area) for a given i instance, a is the total area of the DA segment at a given i instance overlapped by the hexagon, and x is the total DA area at a given i 4.instance.

The challenge presented in this research is motor vehicle collisions rely on associations with additional datasets such as demographics, land use, and infrastructure density, which follow dissimilar aggregate units. These aggregate units, in most cases, follow different scales and boundaries, causing issues with uniformity. In addition, if the administrative boundary chosen is too high in scale, it will diminish its ability to show variation that exists within its unit.

Considering all units of analysis (other than raw point data) use some form of aggregation, the approach described herein aims to address the impacts in relation to the modifiable areal unit problem (MAUP) by introducing a uniform and equal

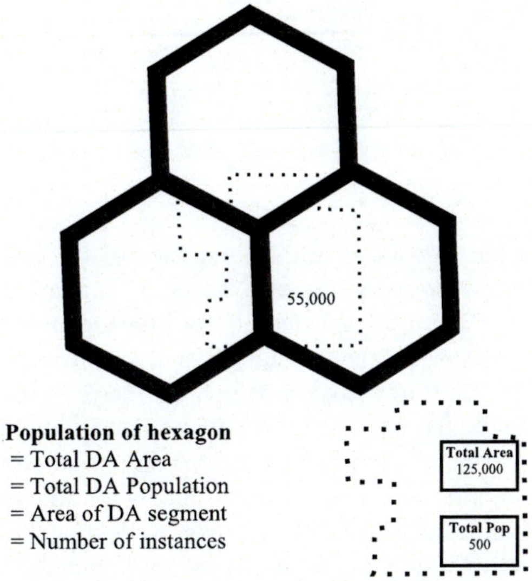

Fig. 4.6 Proportional coverage statistics equation

unit of measure, in this case hexagons. MAUP is the phenomenon that occurs when units of measure are unequal and cause misleading representation of data distributions. The problem is a statistical sampling bias that results from arbitrary, unequal, or dissimilar aggregation units (Dark and Bram 2007).

4.3.3 Spatial Autocorrelation: Getis-Ord Gi Statistic*

Spatial approaches will include the utilization of advanced techniques such as Global Moran's I and Local Getis-Ord Gi* Spatial Autocorrelation testing to determine the level and type of interaction occurring across space both at the region (Global) and neighborhood (Local) level. Testing will evaluate the dependent variable in question, killed or seriously injured (KSI), to understand if the locations in which these events occur are dispersed (value of −1), random in nature (0), or clustered in close proximity to one another (+1) (Getis and Ord 1992). This method of testing allows for an unbiased evaluation to identify cluster intensity and the statistical significance of its observations. KSI datasets will be grouped by their associated regions (Etobicoke, York, Scarborough, and Toronto) to best represent observations that exhibit similar distributions. Evaluations also aim to measure spatial interaction at the system level and compare that at the region level to better understand underlying characteristics.

The calculations associated with Getis-Ord Gi* Statistic allow for examining the level of spatial intensity on a particular unit and compare this to its surrounding neighbors. This approach follows the assumption that units within closer proximity to one another have a higher chance of influence in comparison to values further

$$G_i^* = \frac{\sum_{j=1}^{n} w_{i,j}x_j - \underline{X} \sum_{j=1}^{n} w_{i,j}}{S\sqrt{\frac{\left[n \sum_{j=1}^{n} w_{i,j}^2 - \left(\sum_{j=1}^{n} w_{i,j}\right)^2\right]}{n-1}}}$$

Fig. 4.7 Equation for Getis-Ord Gi* Statistic (Part 1)

apart. As stated in the First Law of Geography by Waldo Tobler, "Everything is related to everything else, but near things are more related than distant things" (1969).

Two important outputs for consideration are derived from this model: the Gi* Statistic and the P value (Bivand and Wong 2018). The Gi* Statistic is outputted in the form of a z-score which can be used to determine the level of clustering of a particular unit and allows for comparison to its spatial neighbors. By measuring the number of standard deviations an observation is away from the mean of the entire sample, the use of z-scores allows for an objective and unbiased approach to measurement. This can be modeled following the equation (see Fig. 4.7), where x_j is the value of a given feature j, w_{ij} represents the spatial weight between feature i and j, and n represents the total number of features within the sample.

Moreover, further declarations are required to complete the calculation (see Fig. 4.8), where represents the mean value of all observations within the test and **S** represents the standard deviation of the observations in question.

Upon determination of whether the spatial interaction of KSIs occurs completely at random, an evaluation of their spatial interaction over time will then be measured to address underlying factors. The outcomes of spatial autocorrelation tests will evaluate the annual characteristics of KSI interaction revealing patterns and trends between 2006 and 2019. Further data modeling will isolate observations by different regions of the city (Etobicoke, York, Scarborough, and Toronto) to obtain a better understanding of how spatial interaction changes through the lens of both space and time.

4.3.4 Model Evaluation and Stepwise Regression

Focusing attention on the clusters that experienced the most spatial autocorrelation, of which are of greatest concern, will allow for an explanatory process that only represents areas of increased concern for injury propensity. For the purpose of this research, only observations where the Getis-Ord Gi* Statistic has a z-score >= ±1.65 will be included in the regression modeling. Additionally, the high volume of variables being tested against KSI occurrence rates will require a modeling approach that can efficiently measure the explained variation that may exist within the dependent variable. Upon identification of the highest influencing variables through a backward stepwise regression model, a forward approach will then be applied using the identified variables in the remainder of model results and evaluations. This approach will allow for a practical evaluation of each variable's prediction strength

Fig. 4.8 Equation for Getis-Ord Gi* Statistic (Part 2)

$$\underline{X} = \frac{\sum_{j=1}^{n} x_j}{n}$$

$$S = \sqrt{\frac{\sum_{j=1}^{n} x_j^2}{n} - (\underline{X})^2}$$

and a comparison within groups to identify how each influencing factor contributes to the prediction of KSI occurrences.

The results of model evaluation will ascertain whether this research can accept or reject its null hypothesis that predictor variables have no effect on influencing KSI occurrences. Backward stepwise linear regression has been identified as an ideal candidate in performing this evaluation. This approach will offer the ability to evaluate a large number of modeled scenarios to create the strongest variable combination to explain the dependent variable (Chun et al. 2016). Backward stepwise modeling utilizes a stepped elimination approach where all variables are added to the model and in each iteration, variables are removed that yield the weakest correlation. It is important to differentiate this approach from forward stepwise modeling as this iteration begins with the strongest predictor variable and continues to add variables to the model that yield the highest increase on model prediction. The challenge, as explained by Hamilton (1987), is the inability for forward stepwise modeling to detect the strongest combination of predictors. This inability can result in variable suppression where some variables are underrepresented due to the increased combination of selections within the model. Backward stepwise modeling offers a more robust alternative which aims to solve for this problem (Shieh 2006).

The final result provides a set of variables that represent the optimal combination of predictors to explain the most variation of the dependent variable. Moreover, the approach takes into account model significance and reliability, allowing for an optimal selection of variables while minimizing the existence of multicollinearity and the likelihood of overfitting (Vaz 2021).

4.4 Results and Discussion

The following section will review the results of this research and discuss potential linkages that may be observed through revelations of statistical modeling and spatial testing. A thorough review of the model results will be evaluated for the backward stepwise regression approach. Discussions will be tailored toward how variable selection contributes to the strength of model predictions and the statistical requirement to meet model assumptions to assure adequate hypothesis testing can occur.

Statistical evaluations will aid in examining potential predictor variables of collisions location resulting in KSI and better indicate the propensity for injury in any given location throughout the City of Toronto. Stepwise linear regression will offer a universally applicable model that can be leveraged by various stakeholders and governments to best support initiatives to minimize the likelihood of KSI and contribute to the academic research community. Predictor variables will be grouped into different themes, (i) environmental variables, (ii) demographic variables, and (iii) infrastructure variables, to best evaluate the level of variation that can be explained by the independent variables (predictors).

Global and Local spatial autocorrelation will be assessed on the locations of KSI occurrences to better understand the underlying characteristics of KSIs within significant clusters. Identification of precise road segments and corridors will be provided to allow local governments, policy makers, and law enforcement to focus resource in these areas to reduce future injury propensity.

4.4.1 Global Autocorrelation and Temporal Characteristics of Injury Incidents

Spatial testing will afford the ability to better understand the level of randomness that may exist over space. This approach will not only consider the spatial interaction of total KSI occurrences but also aims to assess how KSI occurrences change over time. Global spatial autocorrelation tests will follow a queen's contiguity and revealed the City of Toronto had a Global Moran's I value of 0.22 for all KSI occurrences between 2006 and 2019. Findings reveal a moderate level of cluster intensity exists and the spatial interaction of KSI occurrences has a tendency to be driven by underlying factors. When KSI occurrences are broken down by the former municipal regions, it was found that that Toronto yielded the highest level of clustering with a Global Moran's I value of 0.37 (see Fig. 4.9). These results are significantly different in comparison to the other regions across the City of Toronto. Scarborough yielded a slightly higher Moran's I value of 0.09 compared to the remainder of the

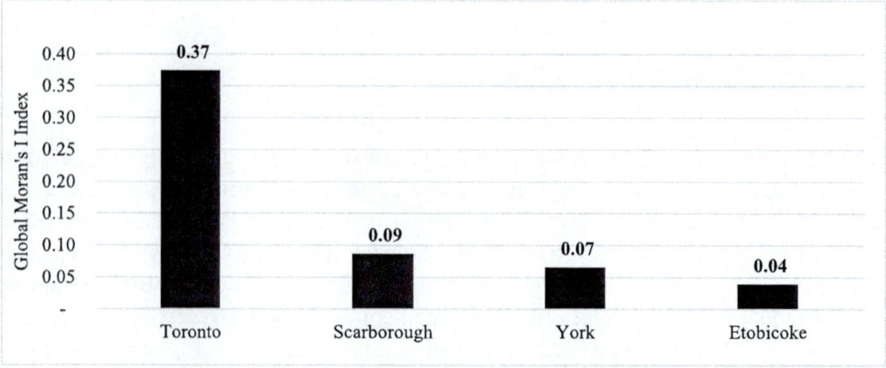

Fig. 4.9 Global spatial autocorrelation of KSI by Toronto Region (2006–2019)

system, but all regions (excluding Toronto) produced an index score between 0.04 and 0.09, making them statistically similar. These observations indicate that global KSI occurrences within these regions are very close to random and not likely to be influenced by their spatial neighbors.

To understand how spatial interaction of KSI occurrences changes overtime, a temporal assessment of Global Moran's I will be evaluated annually from 2006 to 2019. This will then be compared in parallel against annual KSI occurrence volume to understand if the rate of KSI has an effect on the level of spatial interaction. Visual observation of Fig. 4.10 reveals KSI occurrence volumes were trending downward with a 40% reduction in occurrences from 2006 to 2019 (751 vs. 449). Moreover, Global Moran's I values tend to remain at 0.04 on average each year, but an upward trend in clustering was observed in 2011 with a 32% increase compared to the previous year (2011: 0.05 vs. 2010: 0.04). An upward trend continued into 2012 when clustering was observed to be the highest compared to any other year. Interestingly, 2012 also exhibited the highest increase in KSI.

Prior to 2012, an inverse relationship seems to exist. As clustering increases, the volume of KSI occurrences was observed to decrease. However, post-2014, the opposite seems to be observed, i.e., as clustering increases, KSI occurrences also increase at a similar rate. The most notable is the anomaly observed in 2019 where KSI occurrences decreased by −25% (2019: 597 vs. 2018: 449), but clustering was observed to increase the intensity by 109% (2019: 0.05 vs. 2018: 0.03).

4.4.2 *Global Spatial Autocorrelation Trends by Region*

Evaluating the results by the four Toronto regions aims to identify trends within different geographic areas of the city. Comparisons will also be made against the entire system (City of Toronto) to help understand how spatial interaction differs between regions. The City of Toronto was observed to have a positive Moran's I index for all years in KSI occurrences indicating fluctuating level of clustering intensity (see

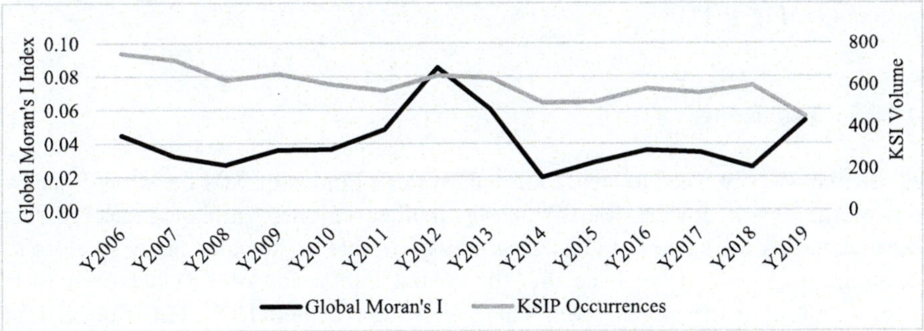

Fig. 4.10 Annual trends of global spatial autocorrelation and KSI volume within the City of Toronto

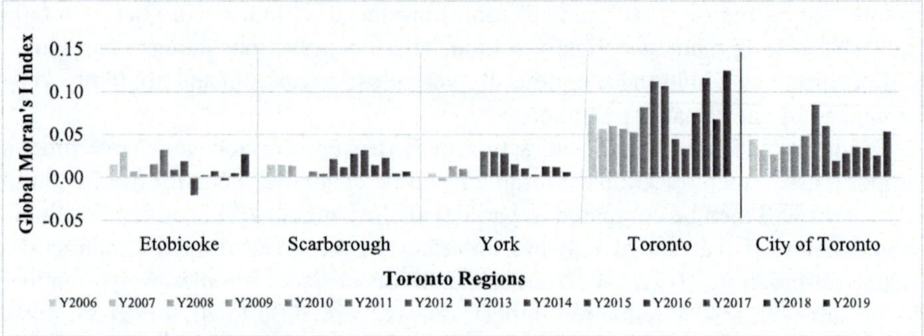

Fig. 4.11 Global spatial autocorrelation by neighborhood and year

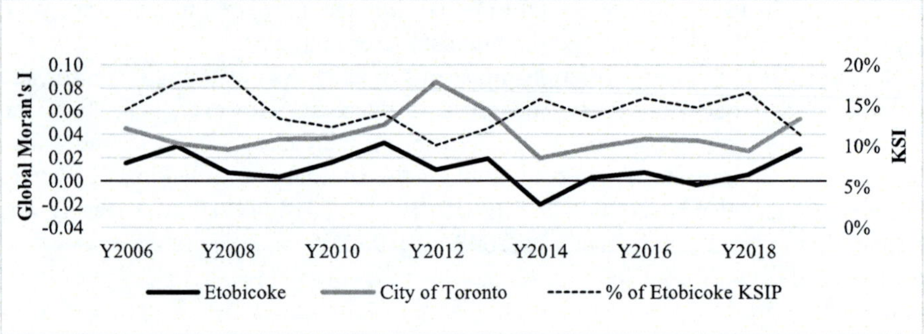

Fig. 4.12 Global spatial autocorrelation within Etobicoke vs. City of Toronto by year

Fig. 4.11). Annual evaluations of Global Moran's I by year and region revealed no notable level of dispersion with the exception of Etobicoke in certain years. Toronto was also observed to have significantly higher levels of cluster intensity compared to neighboring regions. These preliminary findings showcase the importance of evaluating results within various regions of a system to provide additional insights on how the geographic areas may impact the spatial interaction occurring within a system (see Fig. 4.11).

4.4.2.1 Etobicoke

Etobicoke was observed to have a Global Moran's I index of 0.04 for all KSI occurrences, ranking the lowest (fourth) among all other regions. While the index remains positive indicating a level of clustering may exist, due to the value being so close to 0, an assumption can be made that the spatial interaction of KSI occurrences is likely random at the global level. This region represented 15% (1213) of all KSI occurrences within the City of Toronto and had a density of 0.66 KSI per Sq Km each year. Visual observation in Fig. 4.12 reveals spatial autocorrelation patterns within Etobicoke are very similar to that exhibited for the City of Toronto with a mean variance of 0.03 between the two groups. In all years, the rate of spatial interaction was observed to be closer to random within Etobicoke compared to the City of Toronto. Etobicoke also experienced spatial autocorrelation below 0 in 2014 and 2017 indicating some level of dispersion was present.

Prior to 2012, spatial interaction was observed to have a low-strength positive relationship with percent of Etobicoke KSI occurrences at 0.06. Comparatively, results from 2012 to 2019 showcased a strong inverse relationship at −0.61; as clustering intensity increases, KSI occurrence rates were observed to decrease. In 2014, the highest level of dispersion was observed in Etobicoke compared to any other region (see Fig. 4.11). Moreover, 2014 was also observed to have the highest increase in KSIs at 30% from 2013, which was then followed by lower levels of cluster intensity.

4.4.2.2 Scarborough

The Scarborough region was observed to have a Global Moran's I index of 0.09 and was the second highest index of all regions. While spatial interaction in this region remains close to random, it represented a higher level of clustering compared to all other region except Toronto. Scarborough accounted for 25% (2048) of all KSI occurrences with a density of 0.74 KSI per Sq Km each year. Annual trends observed in Fig. 4.13 reveal Scarborough did not experience any levels of dispersion during the observation period. Cluster intensity remained lower in this region compared to the City of Toronto, with the exception of 2014 and 2015 where clustering was 42% and 12% higher, respectively, compared to the system index. When comparing correlation rates between KSI occurrence volume and Global Moran's I Index, a moderate-positive relationship was found at 0.29. This indicates that cluster intensity and percent of KSI occurrences experience increases at similar rates.

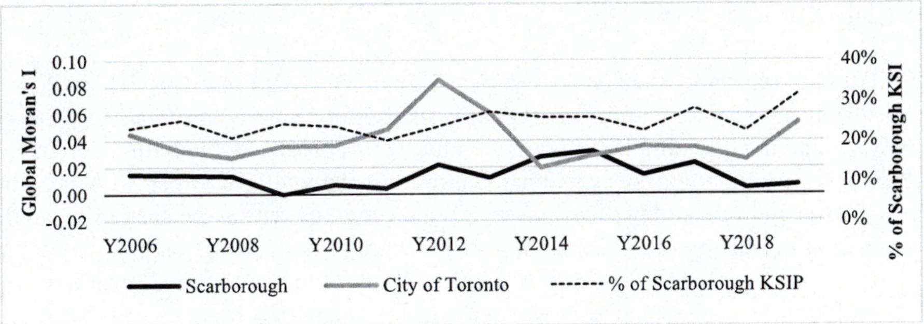

Fig. 4.13 Global spatial autocorrelation within Scarborough vs. City of Toronto by year

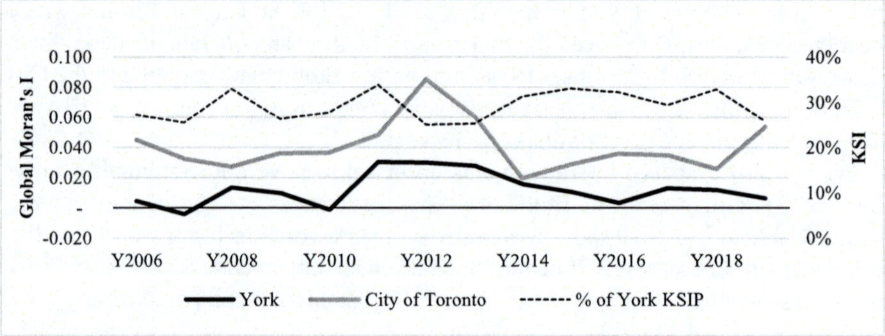

Fig. 4.14 Global spatial autocorrelation within York vs. City of Toronto by year

4.4.2.3 York

The York region was observed to have a Global Moran's I index of 0.07 and ranked second lowest among all other regions in cluster intensity. While the index rate remains close to zero indicating global KSIs occur at random over space, clustering intensity was marginally higher compared to Etobicoke. York represented 29% (2441) of all KSI occurrences within the City of Toronto, making the density approximately 0.75 KSIs per Sq Km each year.

When evaluating annual trends (see Fig. 4.14), it is apparent that York experienced a higher change in clustering intensity compared to the City of Toronto in 2011. This increase was also observed in percent of York KSI which increased by 22% compared to 2010. After 2011, spatial autocorrelation was observed to steadily decline closer to zero by an average rate of −30%, with the exception of a small increase observed in 2017. Correlation between KSI volume and spatial autocorrelation in this region was determined to be −0.43 indicating a medium-strong inverse relationship exists between the two groups. As occurrence volume increased, a decrease in spatial interaction was observed.

4.4.2.4 Toronto

The Toronto region was characterized by a Global Moran's I index of 0.37 indicating spatial interaction within this region is likely not random and tends to cluster among spatial neighbors with similar values. Toronto accounted for over 32% (2651) of all KSI occurrences and had an injury density 1.70 KSI per Sq Km each year. This region ranked highest among all other regions within the system. Spatial interaction in this region is likely more influenced by underlying factors discussed in Sect. 4.4.4. Further evaluations at the local level will reveal the KSI characteristics within these regions and assess where cluster intensity is higher.

Annual trends in Fig. 4.15 show cluster intensity in this region was higher and exhibited very similar trends to those observed within the City of Toronto. This

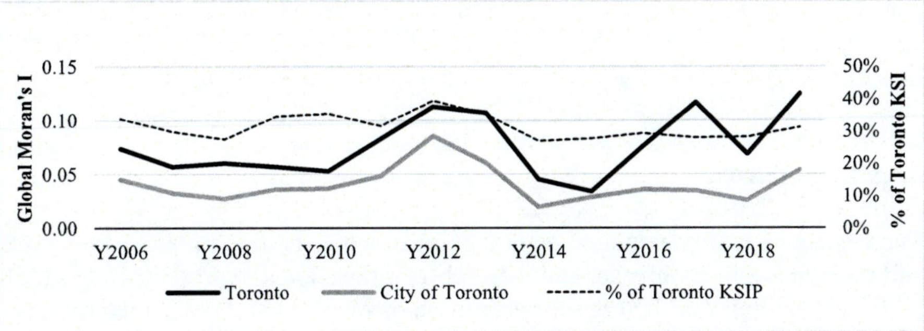

Fig. 4.15 Global spatial autocorrelation within Toronto vs. City of Toronto by year

finding reveals Toronto has a high level of influence on the results for the remainder of the system. This also explains why all regions except Toronto did not present similar annual trends in spatial autocorrelation compared to the system average. While spatial interaction in Toronto is very similar to the system, an increase in clustering was observed in 2016 and 2017 which was found to be significantly higher compared to the city average. Percent of KSI occurrences within Toronto were found to have a correlation rate of 0.29 with spatial autocorrelation. Observations indicate a positive, medium-strength relationship exists between occurrence volume and cluster intensity.

4.4.3 Spatial Clusters and Quantitative Characteristics of KSI by Region

Spatial autocorrelation testing was conducted at the local level for each of the four regions of the City of Toronto. This isolation by region aims to stabilize the variance and distributions of observations, allowing for a more representative assessment of each region. The decision to isolate groups was supported by findings shown in Sect. 4.4.2 which determined Etobicoke, York, and Scarborough exhibited significantly different levels of global spatial interaction (autocorrelation) compared to the Toronto region. To ensure consistency of observations across all region of the system, significant clusters will only be identified when at least two units (hexagons) have a z-score of +/−2.58. This method allows for rejection of the null hypothesis that KSI occurrences are random in nature and do not exhibit clustering or dispersion with the highest confidence level of 99%.

The highest scoring hexagon will then be selected along with its neighboring features to have a total of seven units within each identified cluster. An average z-score of the sample will then be assigned and compared to others for prioritization in future decision-making on injury prevention. Samples with an average z-score of <=+1.63 will be not be considered in this discussion as they exhibit a distribution that is less than 90% confident in rejecting the null hypothesis. Findings will be an invaluable resource for utilization by law enforcement, policy makers, and infrastructure designers to help reduce future injury propensity within these areas.

4.4.3.1 Etobicoke

The Etobicoke region exhibited eight significant clusters, five in North Etobicoke and three in South Etobicoke, and these observations alone represented 21% (258 KSI occurrences) of all KSI occurrences in Etobicoke (1213). The characteristics of KSI occurrences within these clusters revealed 14% of injuries turned fatal (35) and the two most dominant road users injured include automobile operators (74%) and pedestrians (18%). The morning daypart (0600–1159) was overrepresented compared to all other dayparts accounting for 43% (24) followed by the evening daypart (1800–2359) at 25% (14).

The five significant clusters identified in North Etobicoke accounted for 13% (152) of all KSI occurrences within the Etobicoke region. Findings from Getis-Ord Gi* Statistic indicate the northern region of Etobicoke experienced higher levels of spatial interaction compared to the southern region (see Fig. 4.16). Cluster NE1 located near Albion Rd and Finch Ave was observed to have the strongest level of

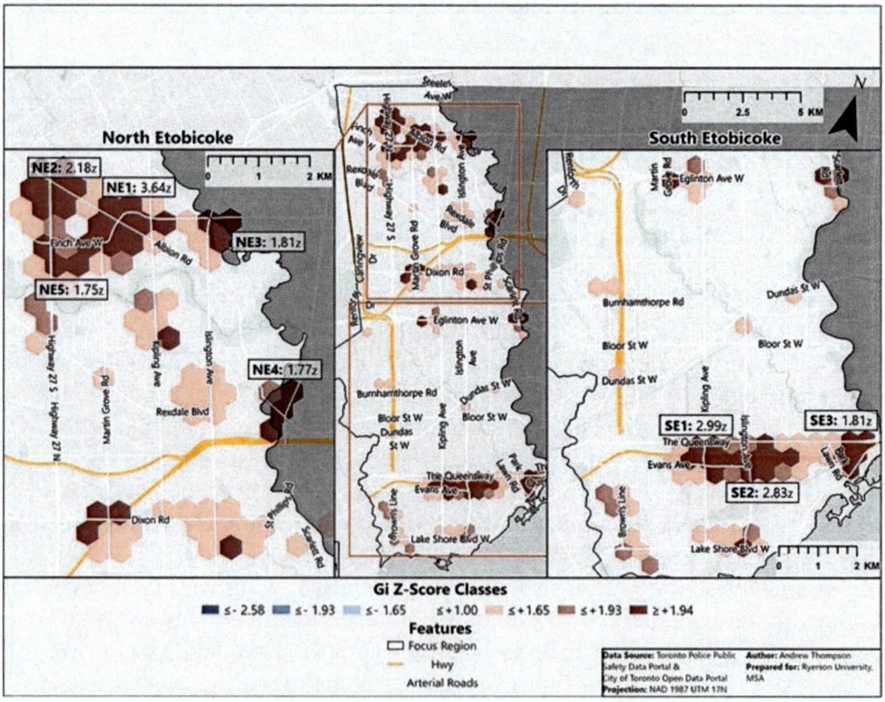

Fig. 4.16 Map of hotspots of KSI collisions in Etobicoke (total between 2006 and 2019)

clustering (see Fig. 4.16) with an average z-score of 3.64. This cluster also represented the highest KSI occurrence rate compared to all other clusters with a total of 49. NE5 was located near Finch Ave and Hwy 27 with a z-score of 1.75 and contained 29 KSI occurrences, and this cluster was characterized by the highest major road density (MRD) with a cumulative length of 4.2 KM.

The southern region of Etobicoke was observed to have three significant clusters accounting for 9% (106) of all KSI occurrences in Etobicoke. These clusters were observed along three major arterial roads (The Queensway, Evans Ave, and the Gardiner Expressway). Cluster SE1 was observed to be the strongest cluster in this region, located near Evans Ave and Kipling Ave exhibiting an average z-score of 2.99. A total of 40 KSI occurrences were observed in this cluster alone and were characterized by having the highest MRD (3.3 KM) among the other two clusters in this region.

4.4.3.2 Scarborough

The Scarborough region exhibited a total of ten significant clusters, six in North Scarborough and four in South Scarborough. Between 2006 and 2019, a total of 2048 KSI occurrences were observed in this area and represented 25% of all KSIs in the City of Toronto. Within the ten identified clusters, there were a total of 406 KSI occurrences, accounting for 20% of all injuries/fatalities within Scarborough. Characteristics of these clusters revealed 15% of KSIs were fatal (60). The two most susceptible road users were automobiles (247, 61%) and pedestrians (130, 32%). Occurrence by time of day reveals 59% of KSIs occur during the afternoon and evening hours between 1200 and 2359.

The six clusters observed in North Scarborough represented a total of 222 KSIs (55%) and were located along three major arterial roads: Finch Ave E, Sheppard Ave, and Ellesmere Rd. The strongest cluster in North Scarborough was NS1 located just north of McCowan Rd and Finch Ave E (see Fig. 4.17) with an average z-score of 2.86 and a total of 34 KSI occurrences. Interestingly, this cluster also exhibited the lowest MRD among all others within Scarborough at 1.9 KM compared to an average of 3.1 KM. NS6 located near Warden Ave and Finch Ave E contained 37 KSI occurrences and exhibited the highest MRD at 3.9 KM. Additional clustering was observed along Sheppard Ave between Warden Ave and Kennedy Rd (NS4) and Ellesmere Rd between Brimley Rd and Markham Rd (NS2 and NS5) and at Victoria Park Ave and Ellesmere Rd (NS3).

The South Scarborough region had four significant instances of clustering and accounted for 184 (9%) of all KSIs in Scarborough. The most significant cluster was observed along Eglinton Ave E between Kennedy Rd and Brimley Rd (SS1; see Fig. 4.17) with an average z-score of 4.93. SS1 represented a total of 40% (73) of all KSIs within the four identified clusters in South Scarborough. Significant clusters were also observed in the following areas: Eglinton Ave E and Victoria Park Ave (SS2), Lawrence Ave E and Warden Ave (SS3), and Lawrence Ave E and Kingston Rd (SS4). The portion of fatalities within these clusters were higher compared to

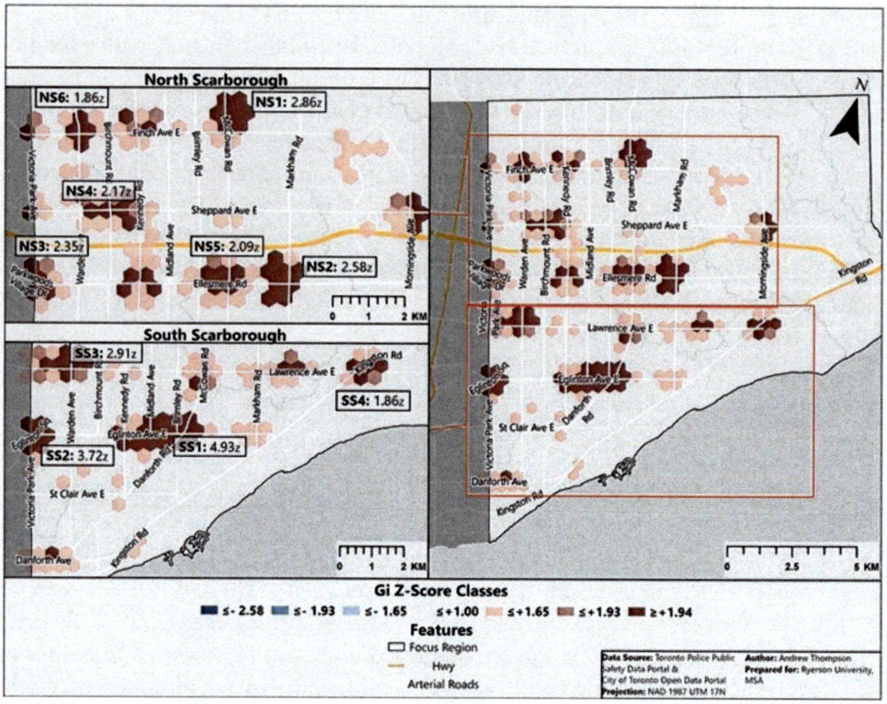

Fig. 4.17 Map of hotspots of KSI collisions in Scarborough (total between 2006 and 2019)

North Scarborough with a total of 45 (25%) of South Scarborough KSIs. Occurrences by time of day revealed an overrepresentation of KSIs during the evening hours of 1800–2359 with 39% (44) of KSIs. These finding may indicate underlying conditions such as infrastructure design, driving behaviors, and demographic characteristics may play a key role on injury propensity and would require additional road safety initiatives to reduce KSIs in the future (Anderson 2006).

4.4.3.3 York

The York region had a total of 2441 KSIs between 2006 and 2019 representing over 29% of all KSIs in the City of Toronto. A total of ten significant clusters were identified within this region accounting for 383 (16%) of all KSIs within the York area. A notable representation of clusters was identified in the south-east region of York where three clusters were observed with an average z-score ranging from 3.88 (Y1) to 2.52 (Y6) (see Fig. 4.18). Interestingly, Y1 located in the corridor of Eglinton Ave E, Weston Rd, and Keele St was observed to have the highest spatial interaction among all other identified clusters. It also represented the highest portion of KSIs compared to any other cluster with a total of 49; findings can likely be explained by a 31% higher MRD density within this area at 3.8 KM compared to a system

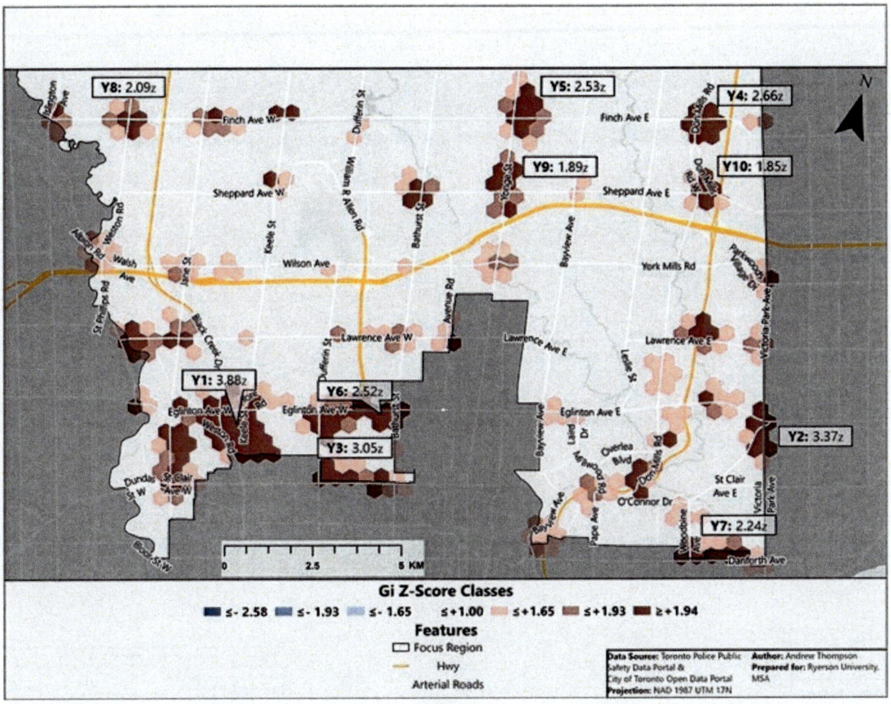

Fig. 4.18 Map of hotspots of KSI collisions in York (total between 2006 and 2019)

average of 2.9 KM for the York clusters. Additional clustering was observed on both Eglinton Ave E (Y6) and St Clair Ave W (Y3) between Dufferin St and Bathurst St, indicating a likely correlation with infrastructure design or driving behaviors within these segments that contribute to an increase in spatial interaction of KSIs.

The north-east region of York revealed four significant clusters, of which two were located in Yonge St (Y5 and Y9) and two along Don Mills Rd (Y4 and 10). A unique pattern was presented with these clusters where the northern clusters (Y4 and Y5) experienced higher levels of spatial interaction compared the southern clusters (Y9 and 10). Further evaluations reveal a 14% higher level of injury propensity in the northern clusters (Y4 and Y5) with an average of 45.5 KSIs per cluster compared to the Y9 and Y10 with an average KSI of 40 per cluster.

4.4.3.4 Toronto

The Toronto region had a total of 2651 KSIs accounting for 32% of all injuries/fatalities within the City of Toronto; interestingly this region had the highest KSIs and the smallest total area. The KSI rate in Toronto was 137% higher at 23.8/Sq Km (between 2006 and 2019) compared to all other regions which exhibited an average rate of 10.03 KSIs/Sq Km. Findings reveal underlying characteristics within this

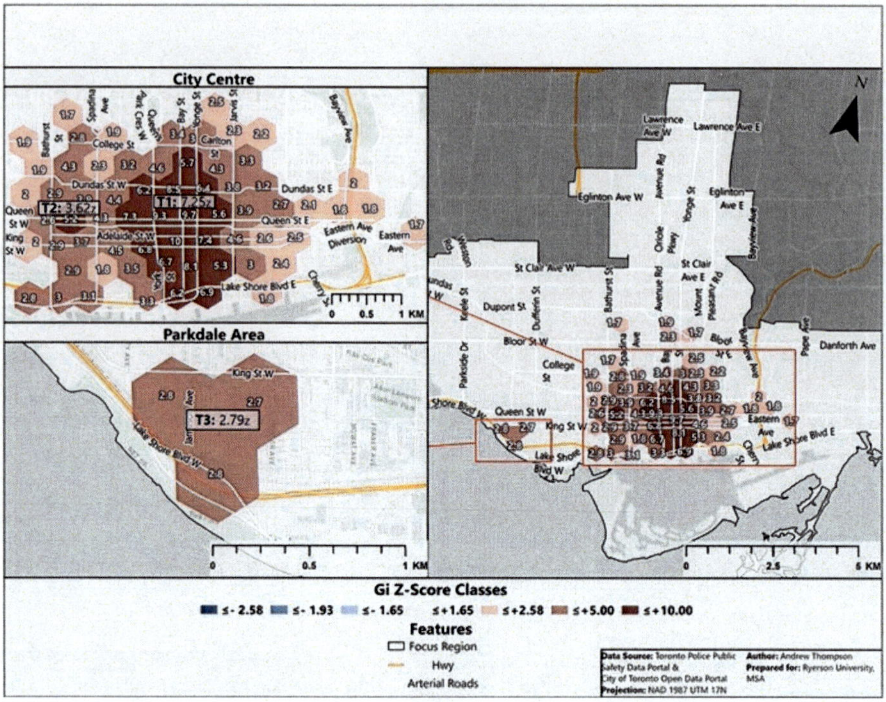

Fig. 4.19 Map of hotspots of KSI collisions in Toronto (total between 2006 and 2019)

region are likely to be significantly different than any other region within the City of Toronto. With this being said, clustering was only present within the city center and an isolated instance in the Parkdale area (see Fig. 4.19). The city center revealed one large mega-cluster of which all hexagons had a z-score higher than 5; this mega-cluster had an average z-score of 7.25 with the highest spatial interaction near the centroid at Bay St and Adelaide St W (T1). Cluster T1 was characterized by a total of 313 KSIs representing 12% of all injuries/fatalities within the Toronto area. Most importantly, this region exhibited a higher portion of injuries related to pedestrians (132, 46%) compared to automobiles (110, 38%).

Cyclists were also overrepresented in this region compared to any other across the City of Toronto with a total of 46 KSIs accounting for 16% of all injuries in Toronto area. Evaluations of occurrences by time of day revealed 40% of KSIs (84) occur in the afternoon between the hours of 1200–1759. During this time, a peak was observed between 1200 and 1259 with 14 KSIs (17% of afternoon occurrences) followed by a second peak between 1400 and 1759 with 62 KSIs (74% of afternoon occurrences). This is likely driven by an increase in infrastructure utilization by working populations during lunch hours and commuter traffic going home from work after 1400. Further safety improvement initiatives may yield a reduction in KSI rates in this area to improve the safety of vulnerable road users through improved infrastructure design, increased bike lanes, and additional pedestrian safety mechanisms (Cinnamon et al. 2011).

On the west end of the city center, an increase in spatial interaction was observed in cluster T2 located along Queen St W and Richmond St W between Bathurst and Spadina Ave KSIs in this area were once again over represented by Cyclists and Pedestrian injuries.

Lastly, an isolated cluster was observed in Parkdale area (T3) located near Lakeshore Blvd and Jameson Ave with an average z-score of 2.79 and 34 KSIs. Majority of occurrences (56%, 15 KSIs) within this cluster were related to automobiles; this is likely driven by the ability of automotive traffic to interchange between Lakeshore Blvd and Gardiner Expressway. Due to the increase in curvatures of these roadways, there may be a correlation with poor visibility or line of sight causing an increase of KSI interaction. Infrastructure designers and law enforcement are recommended to further evaluate underlying conditions within this area to better understand physical characteristics that may increase the likelihood for injury propensity.

4.4.4 Backward Stepwise Regression

Considering all significant spatial clusters have been identified for each region across the City of Toronto, an assessment of their underlying characteristics can now be evaluated and tested. After testing over 1200 variables in a backward stepwise regression model, the outcome identified 13 factors that represented the optimal combination for explaining the highest variation of the dependent variable (KSI occurrences). As depicted in Table 4.1, a variety of factors were combined to best

Table 4.1 Identified predictor variables from SRM by category

Category	Predictor
Land use (3)	Commercial and local
	Commercial and residential
	Institutional and hospitals
Dwelling (3)	Dwelling constructed (2001–2005)
	Households with no bedroom
	% in private household
Infrastructure (2)	Expressway density
	Major/minor arterial density
Education (2)	Postsecondary education in the Philippines
	Science: technologies and technicians
Income (1)	Average income (lone parent economic families)
Journey to work (1)	Commute <15 min
Language (1)	Knowledge of aboriginal language

represent the influencing factors correlated to injury propensity resulting from MVCs. These include land use (three variables), dwelling (three), infrastructure (two), education (two), income (one), journey to work (one), and language (one).

Evaluations of the model summary depicted in Table 4.2 reveal that with the combination of all 13 variables, the model was able to explain 66.2% of the variation in KSI occurrences. In the first iteration of the model, major/minor arterial density was found to explain approximately 47.6% of the variation in KSI occurrences alone. This remains the largest contributing factor and can be expected since KSI occurrences are more likely to occur in areas with higher densities of road infrastructure. In cases where infrastructure has higher density, an assumption must be established that vehicular volumes in these areas have an underlying contribution to increasing the occurrence rates of KSI as well.

The additional 12 variables of the regression model cumulatively explained 18.5% of the variance in KSI occurrences. When evaluating cumulative model results by variable category it was observed that the variance explained in KSI occurrences was 51.2% by infrastructure, 4.4% by land use, 3.8% by dwelling characteristics, 2.1% by journey to work, 1.8% by education, 1.4% by income, and 1.4% by language. Underlying patterns observed through variable selection identify infrastructure, land use, and dwelling to be the top predictor categories indicating a strong unique relationship with the physical environment (Table 4.3).

Further evaluations of the coefficients summary in Table 4.4 reveal majority of predictor variables (nine) had a positive relationship with KSI occurrences. This relationship is indicated by the beta coefficient which represents the magnitude of

Table 4.2 Model summary

Model iterations	R	R square	R square change	Adjusted R square	Std. error of the estimate
Major/minor arterial density	.690	.476	.476	.474	6.269
+ Commercial and residential	.706	.499	.023	.494	6.146
+ Expressway density	.731	.534	.035	.528	5.940
+ Average income (lone parent economic families)	.741	.548	.014	.540	5.863
+ Dwelling constructed (2001–2005)	.752	.565	.017	.555	5.766
+ Commute <15 min	.766	.586	.021	.574	5.639
+ % in private household	.775	.600	.014	.586	5.558
+ Institutional and hospitals	.783	.612	.012	.597	5.485
+ Science technologies/ technicians	.788	.621	.008	.604	5.440
+ Commercial and local	.794	.630	.009	.612	5.386
+ Households with no bedroom	.798	.637	.008	.618	5.344
+ Knowledge of aboriginal language	.807	.651	.014	.631	5.253
+ Postsecondary education in the Philippines	.813	.662	.010	.640	5.189

Table 4.3 Cumulative R square by predictor category

Category	Cumulative R square
Infrastructure	0.512
Land use	0.044
Dwelling	0.038
Journey to work	0.021
Education	0.018
Income	0.014
Language	0.014
Total	0.662

Table 4.4 Coefficients summary

		Unstandardized coefficients				
Category	Variable	B	Lower bound	Upper bound	Sig.	VIF
	KSI total (dependent)	−.29	−1.89	1.32	.73	
Education	Science technologies technicians	.59	.14	1.04	.01	.93
Dwelling	Households with no bedroom	.27	.12	.42	.00	.57
Journey to work	Commute <15 min	.09	.04	.14	.00	.20
Infrastructure	Major/minor arterial density	.01	.01	.01	.00	.54
Infrastructure	Expressway density	.00	.00	.01	.00	.87
Income	Average income (lone parent economic families)	<.001	.00	.00	.00	.56
Land use	Commercial and residential	<.001	.00	.00	.00	.64
Land use	Commercial and local	<.001	.00	.00	.04	.96
Land use	Institutional and hospitals	<.001	.00	.00	.01	.98
Dwelling	Dwelling constructed (2001–2005)	−.09	−.14	−.05	.00	.55
Education	Postsecondary education in the Philippines	−.27	−.48	−.05	.01	.17
Language	Knowledge of aboriginal language	−.48	−.82	−.15	.01	.77
Dwelling	% in private household	−.65	−1.04	−.25	.00	.50

change a variable experiences as KSI increases by 1 unit. If the beta figure is negative, the predictor is deemed to have an inverse relationship with the dependent variable and will decrease as KSI increases. This can be observed most predominantly in predictors related to postsecondary education in the Philippines, knowledge of aboriginal language, and % in private household. The upper and lower bounds of the beta represent the predictions value at a 95% confidence interval. Confidence intervals in this environment report a small range of variation in predictions. Findings suggest model predictors are reliable and repeatable due to the low variance exhibited (see Table 4.4). All predictor variables were observed to be statistically significant predictors ($p < 0.05$) in explaining the variation in KSI occurrences. While predictors were observed to be statistically significant, this was not observed for KSI occurrences (dependent variable); with $p = 0.73$ inferences can confirm that when all other variables are 0, KSI occurrences do not differ significantly from 0.

The assessment of collinearity ensures variable selection within a given model represents a low correlation among other independent predictors and ensures one of the regression modeling assumptions can be satisfied. This is an important consideration to minimize redundancies within the predictor's ability to explain variation within KSI occurrences. For example, if the two variables increase or decrease at the same rate and exhibit similar distributions, this may have an adverse effect on the outcome of model predictions. When multicollinearity exists between multiple variables, it can be detrimental to the statistical accuracy of a regression model (Kaycheng 2015).

Fortunately, collinearity can be measured with statistical modeling, by creating simulation measures and observing the variance inflation factor (VIF) within a given model. As it iterates through each dimension of the model, it introduces new predictor's variables. The variance fluctuations are then measured to determine if the new predictor variable would be redundant. Generally, if VIF values are higher than 10, it can suggest the presence of multicollinearity; for the purpose of this research, a VIF of <10 will satisfy the assumption of multicollinearity. Results indicate multicollinearity is very low and well below the VIF threshold with majority of predictors falling between 1 and 2 VIF. It is important to note that commute <15 min and postsecondary education in the Philippines exhibited the highest VIF values of 4.9 and 5.9, respectively. While these findings do not contradict the assumption of multicollinearity, these predictors exhibited marginally higher similarities compared to others within the model which make them worth noting.

ANOVA also known as analysis of variance is a helpful evaluation in testing the statistical significance of the relationship between the dependent and independent variables within the model. The null hypothesis in this test states that group means are equal and their associated variations do not differ. This evaluation is conducted by comparing the means of more than two groups via an F test to measure the observed variables between groups allowing for a test of the null hypothesis. When significance is $p < 0.05$, then it is safe to reject the null and accept the alternative hypothesis that the observed groups have a difference in variance. When running this test for the model, a determination can be made if variables selected exhibit a difference in variance, and this difference will be confirmed with the F test. Results can also indicate that a robust variety of independent variables were selected when statistical significance is found. In both cases, the ANOVA test confirmed these propositions and rejection of the null hypothesis can me made; variables do have a statistically significant difference in variance where $F(13,200) = 30$, $p < 0.05$ (see Table 4.5).

Table 4.5 ANOVA test results

	Sum of squares	df	Mean square	F	Sig.
Regression	10,527	13	810	30	.000
Residual	5384	200	27		
Total	15,911	213			

The results of model evaluations indicate that all predictors evaluated exhibit a strong relationship with KSI occurrences, the null hypothesis can be rejected, and the alternate can be accepted. Findings confirm that the 13 predictor variables selected have a statistically significant effect on influencing KSI occurrences.

4.5 Conclusions

The approaches exhibited in this research are aimed to be an applicable model for all major municipalities across Canada and other major cities across the word that present similar urbanization patterns to the City of Toronto. The utilization of uniform aggregate units (hexagons) enabled a variety of datasets to be summarized together for an equal and unbiased evaluation. Furthermore, the use of smaller scale hexagon units of approximately 500 m in diameter allowed for an effective approach in identifying specific road corridors that experienced higher levels of spatial interaction compared to others. It is important to note this would not have been possible with arbitrary units such as census tracts or dissemination areas that are better suited to represent residential, commercial, or industrial characteristics that reside within road networks.

Spatial autocorrelation testing at the global and local level utilizing Global Moran's I and Getis-Ord proved to be an effective measurement in understanding how KSI occurrences interact with one another as it revealed a significant difference in spatial interaction within the Toronto region compared to any other region of the city. While findings at the global level did not indicate significant spatial interaction, with the exception of the Toronto region, significant local clusters were observed across all regions of the city. These findings indicate injury propensity is highly impacted by the microenvironment and characteristics such as land use, infrastructure density, and demographics factors are strong predictors. This statement was supported by the stepwise regression model which revealed a statistically significant relationship between the location of KSIs and the surrounding environment. Rejection of the null hypothesis was established that KSI occurrences are not random in nature and likely driven by underlying factors such as environmental attributes, underlying demographics, and the density of infrastructure.

Moreover, temporal evaluation of global Moran's I revealed that while all Global Moran's I values remain close to zero, once again with the exception of the Toronto region, spatial interaction at the global level was somewhat random. The assessments in this research highlighted the relative change year-over-year to provide insights as to how the system experiences change over time. These results will allow for further research to be conducted on historic infrastructure changes, policy implementations, and road safety initiatives to be compared for an evaluation of their effectiveness at reducing injury propensity. It is also important to consider the low number of occurrences being measured in this research, with an average of 601 KSI occurrences per year across the City of Toronto; the sample size remains challenged by inadequate spatial distributions. This inadequacy can be solved with future research focusing on all collision locations which would increase the insights and complexity of the system allowing for improved inferences to be made.

Each year 1.35 million people are killed on roadways across the entire world; this represents a significant number of deaths that each has the potential to be prevented and save lives (World Health Organization 2018).[2] Unfortunately, injury susceptibility is disproportionately impacting vulnerable road users who are at greatest risk. The methods and evaluation techniques outlined in this research aim to support major cities across the globe as an effective and highly robust approach to optimize injury prevention strategies. As the amount of information being recorded continues to increase, more and more can be done to evaluate underlying characteristics of KSIs to strengthen legislation on risk mitigation. Policy makers, infrastructure designers, road safety planners, and law enforcement may find these results of increasing value as observations and revelations presented throughout this research provide data-driven insights on future decision-making. These stakeholders can then benefit from a streamlined approach to improving road safety and enable greater resources to be expended in a manner which affords the greatest impact to society and road safety.

Appendix: Linear Feature Definitions (City of Toronto 2020)

Highway is designated for fast, long-distance travel with restricted access to sustain high speeds.

Highway Transfer Ramp provides for transfer between road and highway and also between highway and highway.

Arterial Road is usually under regional jurisdiction and is fed by collector roads and in some cases is connected to other arterial roads or collector roads via road ramp.

Collector Roads is designated mainly for travel to and from arterial roads with some driveway access. In metro they are usually under local jurisdiction.

Lane is designated mainly for City of Toronto laneways and is usually under local jurisdiction.

Local Road is designated to service driveway access and usually connects to collector roads or other local roads.

Access Road is dedicated to provide access to or within properties such as townhouse complexes, airports etc.

Pending Road is suggested to identify roads with a planned feature code that awaits council approval. This is not requested until the road is assumed and may be delayed for 6 years or more.

Road Ramps (major arterial, minor arterial, collector, other) provide for transfer between two roads.

Busway is a road dedicated for buses only.

[2] https://www.who.int/violence_injury_prevention/road_safety_status/2018/en/

Major Railway is designated for the fast, long-distance, inter-provincial movement of cargo or passenger trains.
Minor Railway is designated for local public transportation and includes aboveground rapid transit corridors and subway lines.
River is a major waterway.
Creek/Tributary is a minor waterway.
Trail is a pedestrian way designated for recreational purposes and can include foot-powered vehicles such as bikes or roller-blades, etc.
Walkway is a designated path primarily for walking.
Hydro Line is an electricity transportation corridor (high voltage).
Major Shoreline is a boundary of a large body of water, e.g., Lake Ontario shoreline.
Minor Shoreline is a boundary of a small body of water such as a pond or reservoir.

References

Anselin L (1995) Local indicators of spatial association – LISA. Geogr Anal 27(2):93–115
Bivand RS, Wong DWS (2018) Comparing implementations of global and local indicators of spatial association. TEST 27(3):716–748
Camilloni L, Farchi S, Chini F, Rossi PG, Borgia P, Guasticchi G (2013) How socioeconomic status influences road traffic injuries and home injuries in Rome. Int J Inj Control Saf Promot 20(2):134–143
Chun Y, Griffith DA, Lee M et al (2016) Eigenvector selection with stepwise regression techniques to construct eigenvector spatial filters. J Geogr Syst 18:67–85
Cinnamon J, Schuurman N, Hameed SM (2011) Pedestrian injury and human behaviour: observing road-rule violations at high-incident intersections. PLoS One 6(6):e21063
Dark SJ, Bram D (2007) The modifiable areal unit problem (MAUP) in physical geography. Prog Phys Geogr Earth Environ 31(5):471–479
Evenson KR, LaJeunesse S, Heiny S (2018) Awareness of vision zero among United States' road safety professionals. Inj Epidemiol 5(1):1–6
Ewing R, Schieber RA, Zegeer CV (2003) Urban sprawl as a risk factor in motor vehicle occupant and pedestrian fatalities. Am J Public Health 93(9):1541–1545
Flahaut B, Mouchart M, Martin ES, Thomas I (2003) The local spatial autocorrelation and the kernel method for identifying black zones: a comparative approach. Accid Anal Prev 35:991–1004
Hamilton D (1987) Sometimes R2>r2yx1+r2yx2. Correlated variables are not always redundant. Am Stat 41(2):129–132. American Statistical Association
Kaycheng S (2015) Multicolinearity in university rankings. High Educ Q 69:158–174
Martin PG, Crandall JR, Pilkey WD (2000) Injury trends of passenger car drivers in frontal crashes in the USA. Accid Anal Prev 32:541–557
Nunn S, Newby W (2014) Landscapes of risk: the geography of fatal traffic collisions in Indiana, 2003 to 2011. Prof Geogr 67(2, 2015):269–281
Ord JK, Getis A (1995) Local spatial autocorrelation statistics: distributional issues and an application. Geogr Anal 27(4):286–306
Pratte D (1998) Road to ruin: road traffic accidents in the developing world. NEXUS 13:46–62
Scheidt PC, Harel Y, Trumble AC, Jones DH, Overpeck MD, Bijur PE (1995) The epidemiology of nonfatal injuries among US children and youth. Am J Public Health 85(7):932–938

Shieh G (2006) Suppression situations in multiple linear regression. Educ Psychol Meas 66(3):435–447
Vaz E (2021) COVID-19 in Toronto: a spatial exploratory analysis. Sustainability 13(2):498
Waller LA (2005) Exploring goodness-of-fit and spatial correlation using components of Tango's index of spatial clustering. Geogr Anal 37:371–382
Witzell J (2019) Physical planning in an era of marketization: conflicting governance perspectives in the Swedish Transport Administration. Eur Plan Stud 27(7):1413–1431

Chapter 5
Socio-spatial Variation in Informal Caregiving to Ontario Seniors

5.1 Introduction

Informal caregiving has an important role to play in the care for the ill, disabled, and elderly members of Canadian society. Many Canadians are already engaged in informal caregiving on a weekly basis, and this trend is expected to continue as a large population cohort, the Baby Boomers, enter their senior and elderly years.

The impacts of this activity are far-reaching. Though unpaid, informal caregiving has an economic impact in the form of caregivers' lost work hours and productivity, and in relation to demand for public and private caregiving services. This has prompted a policy response from both federal and provincial governments in Canada and foreign governments elsewhere. The private sector has also responded by expanding its offerings in formal caregiving services.

Informal caregiving to seniors is often provided within families, but in highly variable ways. Each family improvises its own solution to balance caregiving demands and the family's resources and to determine when outside help is needed. There is a consensus in the literature that informal caregiving is a complex social phenomenon influenced by many factors, with several emerging as being particularly important: the need for care, marital status, and the availability of informal, public, and private care.

This analysis does not dispute the importance of these factors. Its contribution is to reveal the spatial patterning of informal caregiving to Ontario seniors and to explore the strength of the relationships between factors in geographic space. To date, no such analysis has been done for Ontario. Marital status, cultural attitudes toward family caregiving, and the interplay between the availability of informal and formal care, named the Positive Care Law by one research team, are the factors under consideration.

E. Vaz, *Regional and Urban Change and Geographical Information Systems and Science*, Advances in Geographic Information Science,
https://doi.org/10.1007/978-3-031-24731-6_5

The objectives of this study were:

1. To investigate and determine the spatial pattern of informal caregiving to seniors in the study area of Ontario in 2006
2. To identify hotspots of informal caregiving to seniors
3. To understand and determine the sociocultural factors contributing to the spatial pattern using geographically weighted regression

5.2 Literature Review

5.2.1 Definition of Informal Caregiving

The focus of this research will be unpaid care or assistance to seniors as defined in the census of Canada. Seniors are defined as persons aged 65 years and over. Care or assistance refers to help with the activities of daily living (ADL) such as shopping, banking, housework, dressing, bathing, cooking, traveling to and from appointments, and taking medication. This care or assistance may be provided to seniors within or outside the caregiver's household (Statistics Canada 2012a). It does not include volunteer work for a nonprofit or religious organization and a charity or community group or work without pay in the operation of a family farm, business, or professional practice (Statistics Canada 2009a). Unpaid care or assistance is also known as informal or family care and has been defined as care outside the framework of organized, paid professional help (Pong 2008).

5.2.2 Importance of Informal Senior Caregiving

Demographic

In 2011, the population of Canadian Baby Boomers was 9.5 million (Conference Board of Canada 2013, p. 2). That was also the year that the first Boomers, born in 1946, turned 65 and became senior citizens. It is anticipated that in the next three decades, some Boomers will begin to experience chronic health conditions and will require both formal and informal care (Conference Board of Canada 2013, p. 3). For 2001–2031, the average annual growth rate of disabled elderly needing assistance could be 1.9–2.5%, depending on the health status and living arrangements (living with a healthy spouse) of the population in this age group (Statistics Canada 2008, p. 20).

American research indicates that a potential caregiver to care-receiver population ratio of less than 4 to 1 is problematic (AARP 2013). In Canada's most populous province, Ontario, 2011 census figures suggest that the caregiver support ratio from 2025 (when the oldest Boomers will be in their late 70s) to 2045 will fall well below 4 to 1 in virtually all areas of the province (see Fig. 5.1). It is an open question

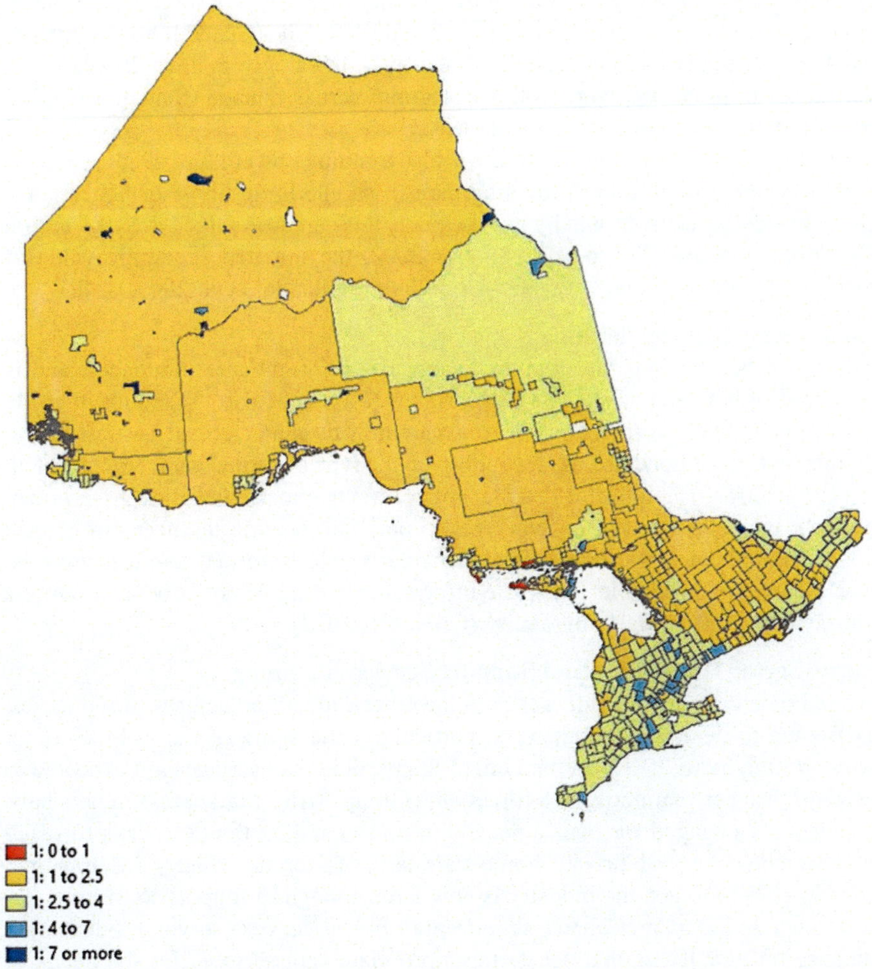

Fig. 5.1 Ratio of projected care receivers (aged 50–64) to potential caregivers (aged 0–49), 2025–2045, Census Subdivisions, Ontario, 2011, Overview

whether the Canadian healthcare system and Canadian families are adequately prepared to provide a suitable mix of care for this segment of the population (Statistics Canada 2003, p. 14).

Economic
In Canadian society, unpaid care to seniors is already a significant activity. In the 2006 census, 4.7 million Canadians aged 15 years or older reported providing unpaid care to seniors, with over one million spending 5–9 h per week, and over 700,000 spending 10 h per week or more. The trend is upward: 16.5% of the population spent time on this activity in 1996, and by 2006 the figure was 18.4% (Statistics Canada 2009b). The 2008–2009 Canadian Community Health Survey found that

one third of Canadians aged 45 or older cared for a senior with a short- or long-term health condition (Statistics Canada 2012b, p. 11). The General Social Survey revealed a 20% increase between 2007 and 2012 in caregivers aged 45 years and over, to 4.5 million caregivers in this age group across Canada (Statistics Canada 2012d, p. 12).

The economic value of informal care is also trending upward. In 1998, Canadians spent over one billion hours providing care to people living in other households, primarily elderly relatives, and by one estimate, this care was valued at $10.3 billion (Statistics Canada 2003, pp. 17–18). For 2009, the imputed economic value of unpaid caregivers in Canada was $25–26 billion (Hollander et al. 2009, p. 48).

Public Sector Service Planning

Rather than being a separate type of healthcare, informal care influences, and is influenced by, the formal care system. Young-Hoon (2006, p. 81) found that the relationship between formal and informal care may be characterized as substitutive, complementary, or nonexistent, depending on the type of formal care. Kitchen et al. (2011, pp. 204–5) noted that despite not receiving much public or government attention, informal care has a "vital role to play" in meeting the needs of elderly people who remain in their homes, particularly where formal services may be unavailable or unaffordable. Adequate informal care can prevent or delay a senior's admission to a residential care facility (CIHI 2010, p. 2).

Private Sector Home Care and Support Services Planning

Private home care and support services have been identified as growth industries, largely due to demographic changes, particularly the aging of the Baby Boomer cohort (Anonymous 2010). In the United States, eldercare services are expected to grow by 6.3% per year until 2018 (Freedonia Group 2015). Market researchers have been actively engaged for over a decade in understanding the best ways to reach potential consumers of private home care services for the elderly (Mathur and Moschis 1999). In fact, the industry is now large enough to support consulting services, such as ElderCareCanada, that do not offer direct care services, but merely guidance on identifying and accessing appropriate care services for the needs of individual elderly care receivers (Fig. 5.2).

5.2.3 Policy Context

International

While there is no international standard for defining unpaid care, the United Nations has made recommendations for national classifications of the activity (United Nations 2008, pp. 166–7 and 244–5). Several countries have pledged to implement a comprehensive system for monitoring informal caregiving, recognizing the crucial role that caregivers play in supporting individuals with health and social care needs. Australia, Ireland, England, and Wales include questions about caregiving in their respective censuses. The United States Census Bureau does not collect data on

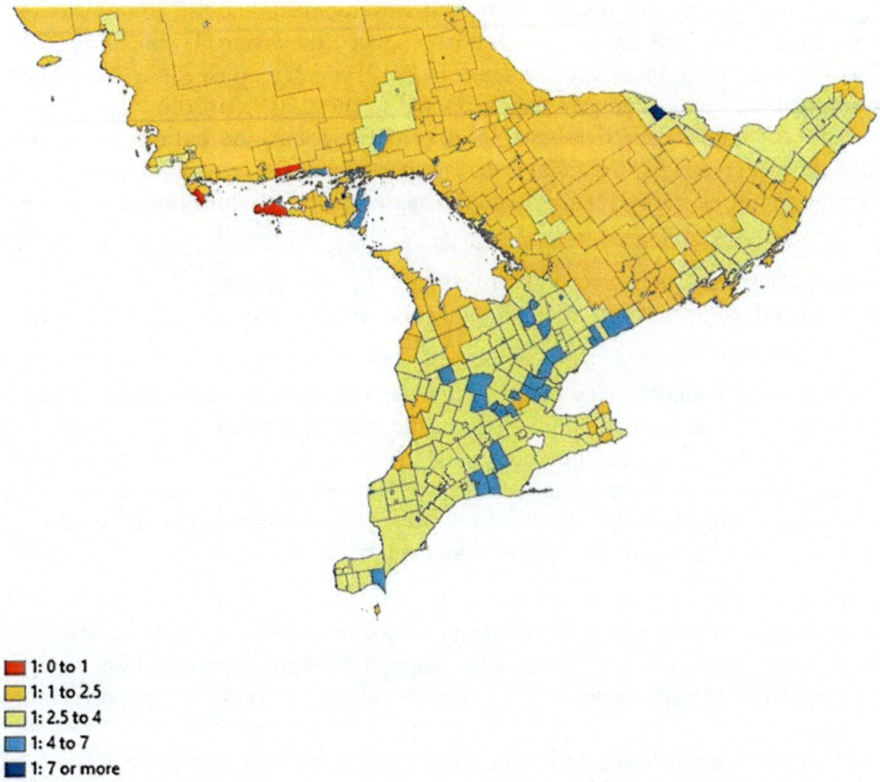

Fig. 5.2 Ratio of projected care receivers (aged 50–64) to potential caregivers (aged 0–49), 2025–2045, Census Subdivisions, Ontario, 2011, Southern Ontario detail

informal caregiving to seniors, but national agencies such as AARP and the National Alliance for Family Caregiving conduct their own quantitative research on the topic. The annual National Survey of Older Americans Act Participants gathers data on caregiving and aging issues from program participants for the purposes of program evaluation and performance measurement (AoA 2015).

There is considerable international research and policy interest in informal caregiving, often in relation to the challenges arising from the aging of the global population. In 2002 the United Nations convened the Second World Assembly on Ageing (the first was in 1982), which produced the Madrid International Plan of Action on Ageing (MIPAA). Over 160 UN Member States adopted the plan. The plan included support for informal caregivers as one of its objectives in recognition of their importance in the continuum of care for the elderly (SWAA 2002, pp. 47–79). Since 2002, dozens of countries have advanced the goals of MIPAA through their own policy initiatives and programs (Global Action on Aging 2015). On an ongoing basis, the United Nations supports research and policy initiatives on aging, caregiving, and related issues in countries worldwide through its Open-Ended Working Group on

Ageing, the Division for Social Policy and Development, and the United Nations Population Fund. The latter organization produced the report *Ageing in the 21st Century: A Celebration and a Challenge* in 2012, which reaffirmed the importance of informal caregivers and caregiver supports (UNFPA 2012, p. 165).

Most developed countries have an array of programs and policies to support informal caregivers, but the models vary greatly. Table 5.1 provides a summary of selected countries' caregiver supports. The supports fall into three broad categories:

1. Flexible work scheduling and leave arrangements
2. Direct supports such as respite care, counseling, and training
3. Financial compensation such as cash allowances and tax credits (Colombo et al. 2011)

Developing countries face many of the same challenges with respect to aging populations and caregiving, although very little is known about the extent and nature of their informal caregiving practices (UNFPA 2012, p. 91). Services and supports to aged persons and their caregivers may be uneven and sometimes nonexistent, which may have the result that there is *more* informal caregiving in developing countries rather than less, as suggested by the Positive Care Law discussed below.

Canada

There has been significant policy interest in unpaid caregiving since the 1980s. This is reflected in the resources that have been directed toward data collection, changes to legislation, income support for caregivers, and programs to support family caregiving.

In addition to the census variables about elder caregiving (see Table 5.2), six of Statistics Canada's General Social Surveys have included questions related to informal care to seniors, and all of the 2012 (cycle 26) General Social Survey focused on this topic (Statistics Canada 2013). The 2008–2009 Canadian Community Health Survey focused on healthy aging and included dozens of questions about caregiving and care-receiving (Statistics Canada n.d.).

The federal *Employment Insurance Act*, through the Compassionate Care Benefits program, entitles eligible caregivers to receive employment insurance benefits in order to partially replace the income lost from periods of unpaid caregiving. The employment legislation of most Canadian provinces allows for some form of job protection for short- and long-term unpaid leave for employees caring for ill relatives, but there is variation in length of leave and the criteria to be met (BC Law Institute 2010, pp. 28–29).

Ontario employment legislation provides for up to 8 weeks of unpaid family caregiver leave per calendar year. This applies to Ontario employees with a specified family member who is experiencing a serious medical condition. The type of care provided may include psychological or emotional support, arranging for care by a third-party provider, or directly providing or participating in the care of the family member (Ontario Ministry of Labour 2015). Additionally, the Ontario Human Rights Commission includes family caregiving in its Duty to Accommodate guidelines (OHRC 2015).

Table 5.1 Summary of caregiver supports in selected countries

	Carers allowance	Allowance for the person being care for	Tax credit	Additional benefits	Paid leave	Unpaid leave	Flexible work arrangements	Training/ education	Respite care	Counselling
Australia	Y	N	N	N	Y	N*	N	Y	Y	Y
Austria	N	Y	N	Y	N	Y	Y	Y	Y	Y
Belgium	Y**	Y	N	N	Y	Y	Y	Y	Y	Y
Canada	Y**	N	Y	Y	Y	Y	N**	Y	Y	Y
Czech Republic	N	Y	N	Y	N	N	Y	Y	Y	Y
Denmark	Y	N	N	N	Y	N	N**	Y	Y	
Finland	Y	N	N	N	Y	N	Y			Y
France	N	Y	Y	Y	Y	Y	Y	Y	Y	Y
Germany	N	Y	Y	Y	N	Y	Y	N	Y	
Hungary	Y	N	N	Y	N	Y	Y	N	N	Y
Ireland	Y	N	Y	Y	N	Y	N		Y	Y
Italy	N	Y	N							
Japan	N	N	N	N	Y	N	Y	Y	N	N
Korea	N	N	N	N	N	N	N**	Y	N	N
Luxembourg	N	Y	Y	Y	N	Y	N**	Y	N	Y
Mexico	N	N	N	N	N	N	N	Y	N	N
Netherlands	Y	Y	N	Y	Y	Y	Y	Y	Y	Y
New Zealand	Y	Y	Y	N	N	N	Y	Y	Y	Y
Norway	Y	Y	N	Y	Y	N	Y	N	N	N
Poland	N	Y	N	N	Y	N	N	N	N	N
Slovak Republic	Y	Y	N					Y	Y	Y
Slovenia	N	N	N	N	Y	N	N**	Y	Y	Y

(continued)

Table 5.1 (continued)

	Carers allowance	Allowance for the person being care for	Tax credit	Additional benefits	Paid leave	Unpaid leave	Flexible work arrangements	Training/ education	Respite care	Counselling
Spain	N	Y	N	Y	Y	Y	N	Y	Y	Y
Sweden	Y	Y	N	Y	Y	N	N	Y	Y	Y
Switzerland	N	N	Y	N	N	N	N**	Y	Y	Y
United Kingdom	Y	Y	N	Y	N	N*	Y	Y	Y	Y
United States	N	Y**	Y	N	N	Y	Y	Y**	Y**	Y**

N*: Leave for only a couple of days for emergency reasons is available
N**: No nationwide policy is available but collective agreements exist
Y**: Not at the national/federal level but available in provinces/states/counties
Source: OECD 2009–10 Questionnaire on Long-term Care Workforce and Financing
Source: Colombo et al. (2011), p. 139

Table 5.2 Census variables related to elder caregiving

	2006	2001	1996
Total population 15 years and over by hours spent providing unpaid care or assistance to seniors – 20% sample data			
No hours	✓	✓	
Less than 5 h	✓	✓	
5–9 h	✓	✓	
10–19 h	✓	✓	
20 h or more (also split male/female)	✓	✓	
Total population 15 years and over by hours spent providing unpaid care to seniors – 20% sample data			✓
No hours			✓
Less than 5 h			✓
5–9 h			✓
10 or more hours (also split male/female)			✓

Source: Statistics Canada

There are various streams of income support available to Canadian caregivers, usually in the form of tax credits or direct payments. The federal government allows a Family Caregiver Tax Credit, dependent and disability credits, and medical expenses deductions. Some of these credits are mirrored in British Columbia's tax system (BC Law Institute 2010, p. 37). Manitoba offers a Primary Caregiver Tax Credit (Government of Manitoba 2014). Nova Scotia has a Caregiver Benefit Program that provides up to $400 per month to caregivers in certain circumstances (Government of Nova Scotia 2014).

Programs to support caregiving across Canada are numerous and varied. A 2003 study found that respite care alone differs significantly from province to province with respect to access, funding, services provided, and user profiles (Health Canada 2003). The federal government provides an online information portal about caregiver support programs, organized by province and territory (Government of Canada 2013). It also launched the Canadian Employers for Caregivers Plan in 2014 to address the issue of caregivers' labor force participation and productivity (Government of Canada 2014). It produced its first report, *When Work and Caregiving Collide: How Employers Can Support Their Employees Who Are Caregivers*, in 2015.

5.2.4 Factors Influencing the Provision of Informal Care

Marital Status

A focus on marital status will reveal how much informal care is provided between senior spouses and how the informal care needs of single seniors are met. Spouses are usually the first option for informal caregiving, in the case of married elders.

Spouse caregivers often do not begin to call on other family members or external sources for support unless caregiving responsibilities outstrip their capacity to meet them. Husbands generally engage in less caregiving than wives and are more likely to ask for help sooner (Palo Stoller and Schroeder Miklowski 2008). Seniors will usually only move in with other relatives if they do not have a spouse or cannot afford to maintain a private residence, although this choice is also influenced by culture (Kaida et al. 2009).

Most Canadians aged 65 years or older live in a private household with a married spouse or common-law partner, especially in their early senior years. As seniors age, they are more likely to live alone or in a collective dwelling such as a care facility or retirement community (Statistics Canada 2012c). Spouses are the most common providers of informal care across all age groups and have the highest rate of caregiving intensity, with a median of 9 h per week (Statistics Canada 2014a).

Among seniors 65 years and older, partially met needs were more prevalent among care recipients whose primarily caregivers were friends or neighbors (17%). By comparison, fewer seniors whose caregiver was the son (12%) or daughter (8%) felt that they had not received all the help they needed. Moreover, seniors without children—and therefore fewer potential family member caregivers—were more likely to have partially met needs, 16% compared with 8% of seniors with four or more children (Statistics Canada 2014b, pp. 6–7).

One quarter of Canadian aged 65 and over lived alone in 2011 (Statistics Canada 2012c, p. 2). Over the age of 85, 36.6% of women and 21.8% of men lived alone (Statistics Canada 2012c, p. 3). Seniors who live alone and require ADL care generally do not receive it from a spouse, but from their adult children (Statistics Canada 2011, p. 1). They are also more likely to use formal services (Chappell 1985). The population of elderly Canadians requiring care but without support from a spouse or adult children was projected by one study to increase by 123% from 2000 to 2030 (Gaymu et al. 2010). The impact of divorce on parent-child relationships may have consequences for the availability of informal care and support provided by adult children to their elderly parents (Statistics Canada 2008, p. 3).

However, the lack of spousal support and the uncertainty of support from adult children do not necessarily mean that single seniors' informal care needs are not met. Twenty-six percent of caregivers in Canada identify their relationship to their care-receiver as "other family member" or "friend, colleague, or neighbor" (Statistics Canada 2012d, p. 19). Research on non-kin carers of older people is limited, but one study found that friends and neighbors provide significant levels of care, varying by geographic proximity, relationship closeness, and the type of care, among other factors (LaPierre and Keating 2013).

Ethno-cultural Background

The extent to which cultural attitudes influence human activity can be difficult to determine. Families and individuals within the same culture can vary greatly in their attitudes as well as their behavior. Canada's cultural diversity, along with intergenerational differences among groups, particularly immigrants, furthers the complexity.

Quantitative research has found that the importance of cultural attitudes toward the provision of informal care is unclear, with different studies finding variations among countries and cultural groups. Using data from the Old Age and Autonomy: the Role of Service Systems and Intergenerational Family Solidarity (OASIS) study, which included variables such as help received from family, welfare state regimes, filial norms, geographic proximity, and emotional solidarity, Lowenstein et al. (2008) compared Norway, England, Germany, Spain, and Israel and found significant variation in the provision of informal care to elders within families. In particular, they found that more family help is provided where family responsibility (as opposed to state responsibility) is valued highly (p. 103). As might be expected, family help was also positively influenced by emotional solidarity, geographic proximity, and the number of adult children.

Culture appears to play a significant role in seniors' choice of living arrangements, and this may extend to caregiving behavior as well. Kaida et al. (2009) found that seniors from familistic cultures (Italian, Chinese, South Asian, and East Indian) were more likely co-reside with family members than those from individualistic cultures (British, German, and Dutch), but this choice was also heavily influenced by marital status and economic constraints.

Social attitudes toward eldercare in general do not necessarily align with specific expectations of care within families, on either the parent or adult child sides. Both general attitudes and specific expectations may be at odds with the actual care provided to elders within families (Peek et al. 2008). One explanation for the difference may be that other considerations supersede the desire or expectation for care within the family. For example, some potential family caregivers may face opportunity costs by engaging in caregiving, such as adult children with high-paying jobs who choose to pay for various types of formal care rather than provide it themselves (Johnson 2008, p. 41). Geographic distance is an ongoing challenge for many informal caregivers (Statistics Canada 2011). However, despite these considerations, adult children form the second-largest group of Canadians engaging in informal care, providing care to 24% of care-receivers (Statistics Canada 2014a, b).

Qualitative research into cultural attitudes toward family caregiving is similarly ambiguous. The title of a 2015 study (Donovan and Williams 2015) into Canadian-Vietnamese views of caregiving clearly delimits one end of the spectrum: "It is like eating, you just do it." However, an analysis of face-to-face interviews with Caucasian-Canadians, Chinese-Canadians, and Hong Kong Chinese found that attitudes toward caregiving are not necessarily linked to caregiving behavior, a finding that is consistent with other research on the topic (Chappell and Funk 2012, pp. 1130–33).

Need for Care Versus the Availability of Care (The Inverse and Positive Care Laws)

The concept of the Inverse Care Law was introduced by Julian Tudor Hart in 1971. Examining the distribution of medical services in England and Wales before and after the establishment of the National Health Service, he found an inverse relationship between need for medical care and availability of service. More and better services were available where they were least needed, and vice versa. This relationship was strongest where medical services were more exposed to market forces, and weakest where public services were most readily available.

Shaw and Dorling (2004) revisited the concept of the Inverse Care Law in relation to the provision of both informal care and medical services in England and Wales. The 2001 census provided two key data points for this study: self-reported data on limiting long-term illness (as an indicator of care need) and hours per week spent providing unpaid assistance (as an indicator of care provision). They found that the provision of informal care was "almost perfectly positively correlated" to the need for care. This relationship was strongest in the north of the country, where the Inverse Care Law applied to medical services, suggesting that people step in to provide care where formal care falls short of local needs. Shaw and Dorling called this phenomenon the Positive Care Law.

Though not described in terms of care "laws," the interplay between informal and formal caregiving has been examined in North America as well. Research in Ontario has shown that relationships between informal and formal care can be characterized as substitutive (as the provision of one form of care increases, the other is reduced), complementary (as one increases, so does the other), or nonexistent (the provision of formal care has no effect on informal care) depending on the type of care provided (Young-Hoon 2006). A Manitoba study found that elderly people use both formal and informal care services according to their need and the availability of those supports, suggesting a complementary relationship between the two types of services (Chappell 1985). American research has found that elderly childless people are *less* likely to use formal care services, despite their having presumably less informal care available to them than elderly parents of adult children. One possible explanation is that adult children act as advocates for their parents in accessing formal care services (Choi 1994). Statistics Canada (2008) analyzed both supply and demand factors influencing the informal care needs of disabled older people to predict future needs and found that while demographic trends may result in a lower supply of family caregivers, the disability rates of older persons may be reduced by improved health outcomes. Nevertheless, pressure on the formal care network can be expected to increase (Expert Group on Home and Community Care 2015, p. 1).

5.2.5 *Need for a Spatial Understanding of Informal Caregiving*

Despite the importance of informal caregiving to national economies and to the quality of life for many families worldwide, there has been very little *geographic* research on the topic. Much of the published research has focused on the economic aspects of unpaid caregiving to seniors, particularly labor force participation, or other considerations such as the impact of caregiving on caregivers' health and well-being. Researchers in Ireland, England, and Wales have produced the few analyses that are geographic enough to include actual maps (Young et al. 2005; Foley 2008). Thus, there is a gap in our understanding of this activity globally, not merely in Canada or Ontario. Because caregiving by its nature requires proximity, it is essential to know where informal care demand falls short of supply in order to plan for the provision of formal care and supports to informal caregivers.

Care-receivers are, by definition, not in robust health. They cannot travel significant distances to receive care; therefore, care must be provided where they are, either in their homes or nearby. Thus, caregiving must be highly localized. Understanding the spatial distribution of informal caregiving and the strength of the relationships between contributing factors will enable better prediction of gaps between care needs and resources. This information can be used in service planning for public care services and market research for private services.

While most Ontarians are not involved in any informal caregiving activity to seniors, many are. On the 2006 census, 1,836,430, 18.5% of the population aged 15 years or more, reported that they had provided such care for at least a few hours the previous week. Over 300,000 Ontarians provided more than 10 h per week. Table 5.3 shows the total persons by weekly hours. The broad extent of reported informal caregiving is consistent with its recognized place in the continuum of care for the elderly in the province.

Most seniors prefer to receive care at home or in their communities (Health Council of Canada 2012, p. 6), but there are many challenges. The Ontario Ministry of Health has a Home First policy regarding care for the elderly (MOH 2015). The goal is to enable Ontario seniors to remain in their own homes as long as possible while having their care needs met by a combination of family caregivers and community supports such as short-term nursing and personal support, adult day programs, assisted living, and respite care. However, current levels of community support are frankly acknowledged to be inadequate (Grant and Church 2015). As the population of Ontario seniors grows, demand is expected to increase but it is not certain whether service levels will rise accordingly. For those who can afford them, private care services are available to fill the gap. For some, their care needs simply go unmet; 15% of respondents in a 2012 caregiving study reported not receiving the care or help they needed (Statistics Canada 2014a, b, p. 1).

The need for a relatively detailed spatial analysis of informal caregiving in Ontario is particularly acute at this time. A recent report noted that demand for home care services in Ontario has sharply risen in the last decade, that the supply of home care services has not kept pace with demand, and that family caregivers feel

Table 5.3 Persons aged 15 years or more engaged in unpaid assistance to seniors, Ontario 2006

Hours per week	Total persons
0	7,975,335
Less than 5	1,110,920
5–9	412,475
10–19	155,925
20 or more	157,310

Source: Statistics Canada

overwhelmed and inadequately supported (Expert Group on Home and Community Care 2015). Because home care services are unevenly distributed and inconsistently funded—"all over the map" in the words of a board member of the Ontario Nurses' Association—access to them has been characterized as a "perverse "postal code lottery'" (Grant and Church 2015). Family caregivers are expected to carry out increasingly complex care tasks such as dressing wounds and changing intravenous medicine bags. This puts senior care-receivers at risk if their needs exceed the informal caregiving resources available to them, a situation that can arise at any time. Recognizing that support for caregivers helps them to provide better care for longer, the Expert Group on Home and Community Care listed increasing supports and services to caregivers as the second of its 16 recommendations. As noted above, these supports must be localized in order to reach the seniors and caregivers who need them.

5.2.6 GIS Applications in the Continuum of Care for the Elderly

GIS may be used in diverse areas of healthcare and social service planning. Its application to these fields can be broadly categorized into analyses of (1) accessibility and utilization, (2) equity and health inequalities, (3) location-allocation and optimal location modeling, (4) epidemiological planning, (5) policy and service planning, and (6) public health and information management (Foley et al. 2009). GIS enables the detection of spatial patterning of a variety of health and social indicators, modeling relationships and processes between determinants of health and their spatial distribution, and the simulation and prediction of various scenarios. In the context of caregiving to seniors, GIS can inform and support choices regarding long-term strategic planning as the population of seniors grows, targeting public education campaigns toward the appropriate audiences, the siting of services such as senior day care centers, or the routing of care workers who provide services in seniors' homes.

There have been a few studies that incorporated GIS into an examination of informal caregiving. An Irish study found that rural areas have higher rates of caring generally, and that intensity of caregiving (more hours per week) was concentrated in older inner-city neighborhoods and the edges of rural counties (Foley 2008). In England and Wales, informal caregiving was found to vary by geography (higher in South Wales and Northeastern England) and by ethnicity (highest among the Bangladeshi and Pakistani populations) and was associated with poor caregiver health (Young et al. 2005). In 2002, Foley reported that there has been no specific research on the use of GIS in informal care planning (p. 82). In a study of caregiving in East Sussex, England, caregiving stakeholders (health service managers, informal caregivers, and care-receivers) found GIS useful for mapping gaps in care demand and supply, and for "spatializing" caregivers' perspectives on service quality by mapping textual analysis results from qualitative research (Foley 2002). In general, however, there has been little spatial analysis of informal caregiving, and the application of GIS to the study of aging issues is far from exhausted (Hirshorn and Stewart 2003, p. 135; Schwanen et al. 2012, pp. 1291–1292). The United Nations has called for further investment into data collection, mapping, and monitoring of all facets of the global population of the elderly (UNFPA 2012, pp. 108–110 and 165).

Emerging technologies designed to enable seniors to remain in their homes longer may provide more opportunities to apply GIS to caregiving and aging issues. These technologies include fall detection and environment sensors, Remote Patient Monitoring and Personal Emergency Response Systems, and "circle of care" social networking apps (Aging in Place Technology Watch 2015). Some of these technologies are already equipped with Global Positioning Systems (GPS) and certainly more will enable geo-tagging in some form. Wearable and ingestible monitoring devices (such as Google Glass and "smart pills") are also in development (Center for Technology and Aging 2014). These technologies will generate large volumes of new and probably geo-referenced data that could be analyzed in GIS to support innovations in policy development and service delivery. Some researchers have already begun such analyses (Shoval et al. 2010).

5.3 Study Area

The study area is the Canadian province of Ontario. As depicted in Fig. 5.3, it is bordered on the east by the province of Quebec, on the west by the province of Manitoba, on the north by Hudson's Bay, and on the south by the Great Lakes-St Lawrence River chain, which forms a natural border with the United States. Ontario is the second-largest province by area with 1,076,395 square kilometers.

In 2011, the population was 12,851,821, making Ontario the most populous Canadian province. Originally inhabited by many groups of Aboriginal Peoples, since the late eighteenth century, the province has experienced many waves of immigration from all continents and from elsewhere in Canada. However, Ontarians with British Isles ancestry are still the largest single ancestral group. A notable exception is the city of Toronto, the provincial capital, where two fifths of the population are visible minorities. Northern Ontario and the cities of Sudbury and Ottawa are home to substantial populations of Franco-Ontarians.

Most of the Ontario population can be found in Southern Ontario, in the province's portion of the Windsor-Quebec corridor. The Greater Toronto/Golden Horseshoe agglomeration on the shore of Lake Ontario accounts for about one third of the total provincial population. Other major population centers include Ottawa (the national capital), Kitchener-Waterloo, and London.

Ontario's economy is very diverse. It includes the agriculture, resource extraction, financial services, telecommunications, technology, media, arts, and entertainment industries. Proximity to major American markets is a major economic advantage for the province (Encyclopedia Britannica 2015).

Fig. 5.3 Map of Ontario

5.4 Methods and Data

5.4.1 Methods

Spatial Autocorrelation

Spatial autocorrelation is a statistical technique used to identify geographic patterns, make spatial predictions, and detect deviations (Griffith 1987, p. 4). It is primarily concerned with similarity among attributes and similarity among locations (Goodchild 1986, p. 11). Its most significant result is an index of similarity.

Spatial autocorrelation can produce both global and local measures. Global measures determine *whether* there is a spatial pattern and local measures determine *where* there is a spatial pattern. Global measures create one value for the entire dataset. This suggests that the same process occurs over the whole geographic area, which may not actually be the case. To address this problem, local measures may be used. These create a value for each unit of observation and may reveal that different patterns and process are at work in different parts of the study area.

Global Moran's I

A commonly used global measure of spatial autocorrelation is Global Moran's Index, also known as Moran's *I*. Moran's *I* is structured so that a positive result between 0 and 1 means nearby areas are similar, a negative result between 0 and −1 means they are dissimilar, and a value close to zero means the pattern is random and independent.

The final number is the result of:

1. Calculating the mean and variance for the attribute being analyzed from the feature values
2. Subtracting the mean from each feature value to create a deviation from the mean
3. Multiplying the deviation values for all neighboring features (within a distance specified by the analyst) to create a value called the cross-product
4. Summing the cross-products to form the numerator of Moran's *I* (ESRI 2015)

Below is the formula for Global Moran's *I*:

$$I = \frac{\sum_{i=1}^{n}\sum_{j=1}^{n} w_{ij}\left(z_i - \overline{z}\right)\left(z_j - \overline{z}\right)}{s_z^2 \sum_{i=1}^{n}\sum_{j=1}^{n} w_{ij}}$$

Source: Neiman (2007).

Moran's *I* also produces an Expected Index value, which is compared with the Observed Index values to compute a z-score and a *p*-value. These determine whether the difference is statistically significant.

Local Moran's I

Local Moran's *I* is the same as Global Moran's I, but as its name suggests, it is a local measure. It creates a value for each unit of observation. It is also known as the Anselin Moran's *I* and is one of the best-known Local Indicators of Spatial Association (LISA).

Below is the formula for Local Moran's *I*, with the subscript *i* noting the location:

$$I_i = \frac{\sum_{j=1}^{n} w_{ij}\left(z_i - \overline{z}\right)\left(z_j - \overline{z}\right)}{s_Z^{\,2} \sum_{j=1}^{n} w_{ij}}$$

Source: Neiman (2007).

With Local Moran's *I*, statistically significant z-scores, both positive and negative, are used to further classify the features into clusters and outliers. Each feature's values is compared to the mean for the local (as defined by the analyst) area. For outliers—features with strong negative z-scores—values higher than the local mean are classified as High-Low, and those with values lower than the local mean are classified as Low-High. For clusters—values with strong positive z-scores—the local mean is compared with the global mean. Where the local mean is higher than the global mean, the feature is classified as a High-High cluster, and where it is lower, the cluster is classified as a Low-Low cluster (Rosenshein 2011). Outliers are features with high or low values surrounded by dissimilar features, and clusters are feature with high or low values surrounded by similar features, as detailed in Table 5.4.

Local Getis Ord-Gi* (Hotspot Analysis)

Local Getis-Ord Gi* is another useful local spatial association statistic but not strictly a LISA (Anselin 1995, p. 101). It is a measure of association between a single weighted point or concentration of weighted points and all other weighted points within a given radial distance from the original point or concentration of points (Getis and Ord 1992, p. 190). It produces z-scores and p-values which indicate the degree of clustering among points. z-scores must be statistically significant to be meaningful. A higher z-score indicates a clustering of high values (a hotspot). A negative z-score indicates a clustering of low values (a coldspot). Proper use of this statistic requires at least 30 units of observation.

The Local Getis-Ord Gi* maybe expressed as follows:

$$G_i(d) = \frac{\sum_j w_{ij} x_j}{\sum_j x_j}$$

Source: Briggs (2010).

Table 5.4 Definitions of clusters and outliers resulting from Local Moran's *I* analysis

Cluster or outlier type	Definition
HH—High-High	Cluster of high values
LL—Low-Low	Cluster of low values
HL—High-Low	Outlier: a high value surrounded by low values
LH—Low-High	Outlier: a low value surrounded by high values

Source: Adapted from Scott (2008)

Geographically Weighted Regression

Regression is a common statistical analysis technique. It is used to assess the relationship between one dependent variable and one (or more, in the case of multiple regression) independent variable. The classic regression model is a global model. It produces one set of results for the entire study area that does not account for variation across geographic space. For this reason, it is a poor *spatial* analysis technique.

Geographically weighted regression (GWR) is an alternative that does account for spatial variation. The GWR technique conducts "mini" regression analyses on one subset of data, one location at a time.

The difference between GWR and classic regression is apparent from their respective formulas:

$$\text{Classic} \quad y = \beta o + \beta 1\, x1 + e$$

$$\text{GWR} \quad y(u,v) = \beta o(u,v) + \beta 1(u,v)\, x1 + e(u,v)$$

[where (u,v) represents the location of the observation]
Source: Van Loon (2008).

GWR fits a regression model—a "kernel" around one observation (regression point) and those that are nearby (data points). Each data point is weighted according to its distance from the regression point. The closer the data point, the more heavily it is weighed. It has a maximum weight if it shares a location with the regression point.

An important consideration in using GWR is determining which observations should be included as neighbors to the observation under analysis. This determination is reflected in the choice of kernel type and kernel size.

There are two types of kernels: fixed and adaptive. "Fixed" means that the kernel is the same size at every data point and that the number of observations will vary while the area they represent—the distance from the regression point—will remain constant. "Adaptive" means that the kernel size changes according to the density of data and that the number of observations will remain constant despite changes in the size of the area represented. Figures 5.1 and 5.2 depict the difference between kernel types.

Fixed kernels are most appropriate when there are a lot of observations distributed relatively evenly across space. Adaptive kernels are best suited to cases where the distribution of observations varies. The choice of kernel size is guided by the data and the requirements of the analysis. Kernel size affects the outputs and therefore the quality of the results. A kernel too large will increase bias (the inclusion of observations that are not part of the spatial group). A kernel too small will increase variance (not enough data points in the model to draw worthwhile conclusions). Researchers must find an optimal balance between bias and variance. The model evaluation methods cross-validation (CV) and corrected Akaike information criterion (AICc) may be used to choose an appropriate kernel size.

The primary advantages of GWR are that it can detect spatial heterogeneity and produces results that are fine-grained, essentially *local* statistics. The results are geo-coded, making them "GIS-friendly" and well-suited to being mapped. The significant disadvantages of GWR are that it is not effective at analyzing relationships at a small scale, as there is too much bias, and, as with classic regression, it cannot rectify the problems of a poorly specified model.

GWR has been used on a very wide variety of topics in many disciplines, including rates of conversion from conventional to organic farming across the United States; stray cat populations in Auckland, New Zealand; school performance in the United Kingdom; forecasting transit ridership in Madrid, Spain; and demographics and tobacco outlet density in Polk County, Iowa. This diversity is an indicator of the usefulness of GWR in exploring spatial processes (Figs. 5.4 and 5.5).

5.4.2 Data

Data for Ontario was drawn from two sources, the 2006 census of Canada and the 2009 Canadian Community Health Survey. A list of the variables and their associated census and survey questions is presented in Appendix A. Census data was unavailable for some uninhabited parts of Ontario, large provincial parks, and certain First Nations communities. These areas were not included in the analysis and appear as blank spaces on the maps. For some areas of the province, the count for the health variables of interest was zero, and so they also appear blank.

To account for geographic variation in population characteristics and caregiving activity, each census variable was normalized using the following formula:

$$\frac{\text{Variable}}{2006\,\text{population}} \times 10000 = \text{count per } 10000 \text{ population}$$

Health survey variables were normalized in the same way, using number of respondents in place of population count as the denominator.

From the census, there were two major variables of interest: (1) hours spent per week on unpaid caregiving or assistance to seniors and (2) language most often spoken at home. Both were normalized. Census subdivisions (CSDs), of which there were 527 in Ontario in 2006, were chosen as the unit of analysis. This level of aggregation was chosen because it is assumed to capture both care need and caregiving behavior within each geographic area, and because the number of CSDs was suitable for analysis and mapping in a study area as large as Ontario. Additionally, it was necessary to aggregate the health survey data to the CSD level, and so matching the census data to that level facilitated the creation of more readily comparable maps and results.

The census language variable was used as a proxy for ethno-cultural origin to create two new variables, INDIV for more "individualistic" cultures and "FAMIL" for more "familistic" ones. English and Northern European languages formed the individualistic group and all other languages formed the familistic group. Appendix B has a list of all the languages used in the creation of these variables.

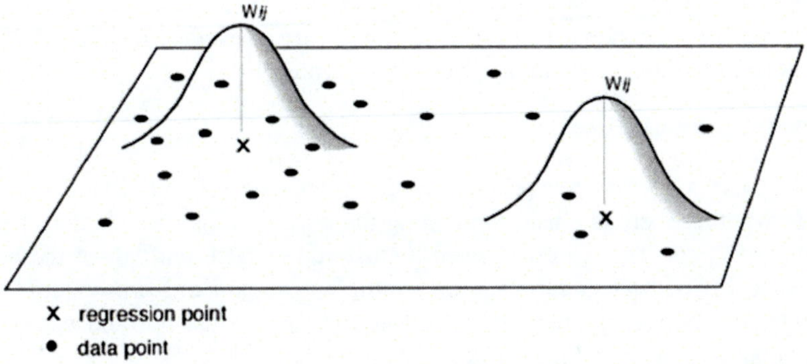

Fig. 5.4 A fixed kernel weighting scheme

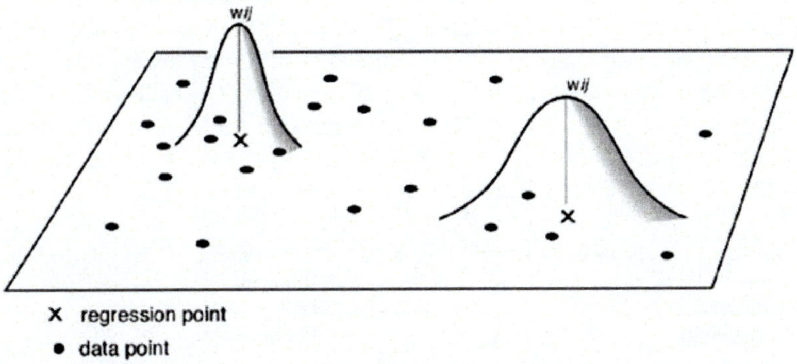

Fig. 5.5 A spatially adaptive weighting scheme. (Source: Fowler 2011)

The categorization of cultures by language is admittedly crude and subjective. For the purposes of this analysis, broad categories of culture were used in order to examine the influence of culture on caregiving behavior. Drawing on the work of Mitchell (2001, pp. 393–4) and Kaida et al. (2009), "Western" nontraditional cultures such as Canada's mainstream, heavily Anglo-American one, are assumed to value individualism and self-reliance. Traditional cultures such as those of Canada's First Nations, Southern Europe, Asia, Africa, and Latin America are assumed to value filial piety and intergenerational support. A further assumption is that these cultural attitudes influence actual caregiving behavior within families, namely, that people from nontraditional cultures will engage in less caregiving and those from traditional ones will do more, all other things being equal.

From the Canadian Community Health Survey, the selected variables were related to:

- Age
- Marital status
- Living arrangements (with spouse, alone, other relatives)

- Needing help with the activities of daily living
- Whether the respondent had received home care assistance from a nongovernment source (family, friend, neighbor, or volunteer)

Data for seniors (persons aged over 65 years at the time of survey) was extracted from the main health survey dataset. After normalization, Global Moran's I and Getis-Ord Gi* analyses were performed on the activities of daily living variable to identify whether there is clustering among the observations and whether there are hot and coldspots. The results showed a clustered pattern, with all of the hot and warm spots in Southern Ontario and the southernmost parts of Northeastern Ontario (Sudbury and Nipissing areas). Census subdivisions in the hot and warm spots, 261 in total, were selected for further analysis.

As noted in Table 5.5, the census variables were used in spatial autocorrelation and hotspot analysis to assess spatial patterning, and geographically weighted regression (GWR) to explore the relationship between culture and rates of caregiving. The health survey data was used in GWR to explore the relationship between marital status and caregiving, and whether the Positive Care Law applies in Ontario. Each of the hours-per-week variables was analyzed using Global Moran's *I*, Anselin Local Moran's *I*, and Getis-Ord Gi* hotspot analysis in ArcMap 10, with the following inputs:

Conceptualization of spatial relationships:	Fixed distance
Distance method:	Euclidean
Row standardization:	Yes
Distance threshold:	1000 km

All of the geographically weighted regression models were conducted in ArcMap 10, with the following inputs:

Kernel type:	Fixed
Bandwidth method:	AICc

5.5 Results and Discussion

5.5.1 Results

The objectives of this analysis are:

1. To investigate and determine the spatial pattern of informal caregiving to seniors in the study area of Ontario in 2006
2. To identify hotspots of informal caregiving to seniors

Table 5.5 Methods and data

Objective	Question	Variables	Method
To investigate and determine the spatial pattern of informal caregiving to seniors	Is there a spatial pattern to informal caregiving to seniors? Are there clusters in different parts of the province?	**2006 Census Question 33(c)** Less than 5 h/week 5–9 h/week 10–19 h/week 20 h or more/week	Global and Local Moran's *I*
To identify hotspots of informal caregiving to seniors	Where are the hotspots of informal caregiving to seniors?	**2006 Census Question 33c**	Local Getis Ord-Gi* (hotspot analysis)
To understand and determine the sociocultural factors contributing to the spatial pattern, per models described below	What influences the level of informal caregiving to seniors in communities?		
Model 1 (marital status)	Is there a relationship between marital status and the level of informal caregiving? Do married seniors received more or less care than single ones?	**2009 Canadian Community Health Survey** **Dependent variable:** HMC_11—has received nongovernment home care **Independent variables:** DHH_MS (married/not married) DHHDLVG (with spouse or partner/unattached, living alone)	Geographically weighted regression
Model 2 (cultural background)	Does ethnic or cultural background influence the level of informal caregiving to seniors?	**Dependent variable:** **2006 Census Question 33c** **Independent variable:** **2006 Census Question 15 (a)** Language most often spoken at home	Geographically weighted regression

(continued)

Table 5.5 (continued)

Objective	Question	Variables	Method
Model 3 (Positive Care Law)	Is there a relationship between the provision of care and the need for care in communities? Does the Positive Care Law apply to eldercare in Ontario?	**2009 Canadian Community Health Survey** **Dependent variable:** HMC_11—has received nongovernment home care **Independent variable:** ADLF6R—need for help with activities of daily living	Geographically weighted regression

3. To understand and determine the sociocultural factors contributing to the spatial pattern using geographically weighted regression

Table 5.6 details the summary statistics for the reported hours per week on the 2006 census. Curiously, the mean drops as the hours per week rise and then rises again for the highest hours-per-week variable, 20 or more hours. Table 5.7 summarizes the pertinent data from the health survey, showing that 1180 out of 4491 respondents, or 26%, need ADL help. Of those, 463 or 39% had received care from a nongovernment, informal source such as a family member or friend.

1. To investigate and determine the spatial pattern of informal caregiving to seniors in the study area of Ontario in 2006

Table 5.8 shows the results of the Global Moran's I analysis. The less-than-five variable was identified as random, while the three remaining variables were found to be clustered.

2. To identify hotspots of informal caregiving to seniors

Figure 5.6 depicts the results of the Getis-Ord Gi* analysis for the less-than-five variable. It shows clear hotspots in the Northeastern and Eastern parts of Ontario, and a definite warm spot running the length of the southernmost part of the province. Cool spots are found in the Northwestern region and in the Northern parts of the Southern regions, such as the Bruce Peninsula and Muskoka areas. Figure 5.7 shows the results of the Anselin Local Moran's I for the less-than-five variable. Nearly the entire province was determined to be not significant, with a handful of outliers in the North.

Figure 5.8 depicts the results of the Getis-Ord Gi* analysis for the five-to-nine variable. It shows a clear South-to-North, cold-to-hot pattern, with the hottest spot in the extreme Northwest. Figure 5.9 shows the results of the Anselin Local Moran's I for the five-to-nine variable, which include High-High clusters in the Rainy River and Lake of the Woods parts of the Northwestern region and Low-Low clusters and

Table 5.6 Summary statistics for caregiving hours per week (rate per 10,000 persons), Ontario 2006

Variable	Minimum	Maximum	Mean	Standard deviation
Less than 5 h	0	2616	974	271
5–9 h	0	1887	406	481
10–19 h	0	1220	157	180
20 or more hours	0	2128	187	218

Table 5.7 Summary of health survey data, Ontario 2009

Total senior respondents	Needing ADL help	Received nongovernment help	Married or common law	Single, widowed, or divorced	Living with spouse or partner	Living alone
4491	1180	463	2250	2232	2016	2079

High-Low outliers scattered throughout Southern Ontario and the Southern part of the North, near Sudbury and North Bay.

Figure 5.10 depicts the results of the Getis-Ord Gi* and Fig. 5.11 shows the results of the Anselin Local Moran's I analysis for the ten-to-nineteen variable. These maps show a similar but weaker pattern than those for the five-to-nine variable: The South is cool, and the North is hot, with a transition from South to North around Sudbury and North Bay. However, there is less clustering overall for the ten-to-nineteen variable.

Figure 5.12 depicts the results of the Getis-Ord Gi* analysis for the twenty-or-more variable. This map shows a similar South-to-North, cold-to-hot pattern as the other maps, but with the hottest spots only in the Rainy River and Lake of the Woods areas and small communities further north. The South is not quite so cold as in the other maps, but it is noteworthy that the coldest parts are in the most populous areas of the southernmost strip along the St Lawrence River and Great Lakes, and the Sudbury/North Bay area. Figure 5.13 has the results of the Anselin Local Moran's I for the twenty-or-more variable, which show High-High clustering in small remote Northern communities and Low-Low clustering scattered throughout the South, particularly in the Kitchener-Waterloo-Guelph area, with most areas of the province deemed not significant by the analysis.

3. To understand and determine the sociocultural factors contributing to the spatial pattern using geographically weighted regression

Model 1 (marital status)—Is there a relationship between marital status and the level of informal caregiving? Do married or partnered seniors receive more or less care than single ones?

Table 5.9 details the R-squared results and interpretation for each model. The low values for R-squared show that the models do not adequately explain the relationship between marital status and receipt of informal care.

Table 5.8 Global Moran's *I* results—hours-per-week census caregiving variables

Variable	Global Moran's *I*	z-score	p-value	Pattern
Less than 5 h	−0.003700	−1.221808	0.221780	Random
5–9 h	0.036377	26.066063	0.000000	Clustered
10–19 h	0.006575	5.801988	0.000000	Clustered
20 or more hours	0.030328	22.249850	0.000000	Clustered

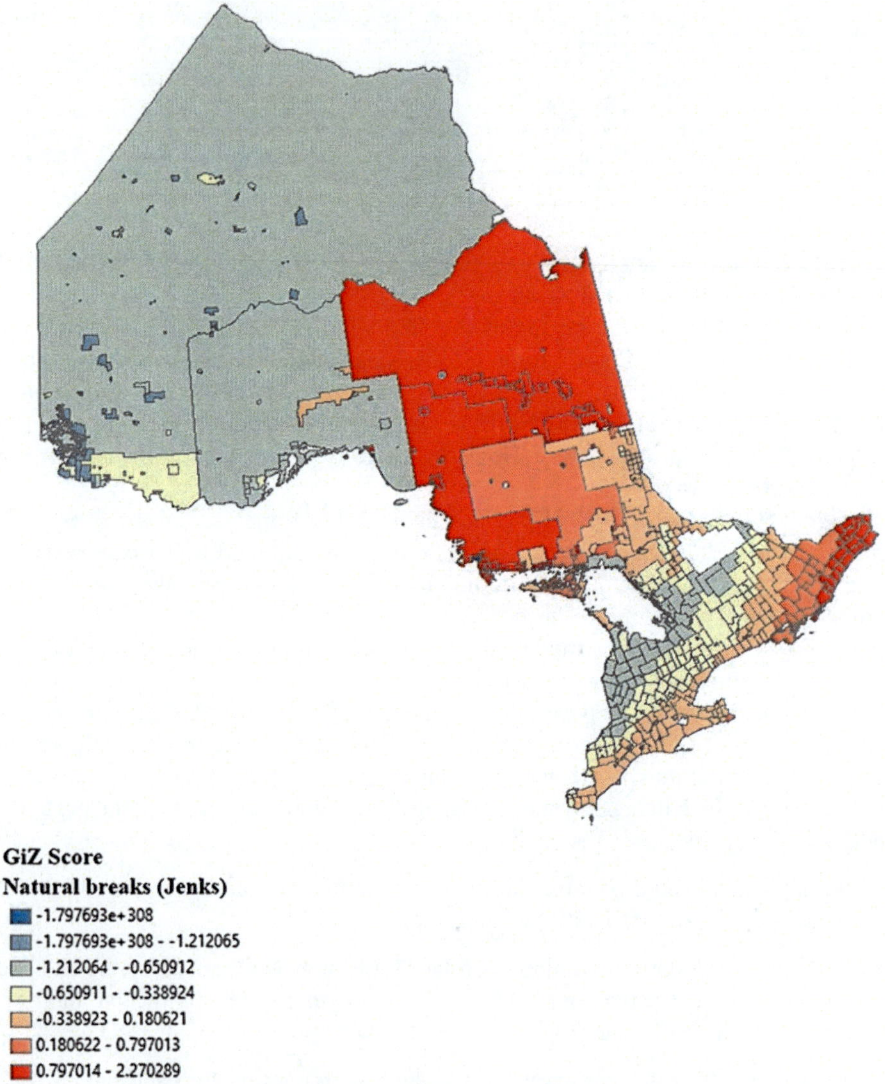

Fig. 5.6 Map of hotspots—less than 5 h of caregiving per week

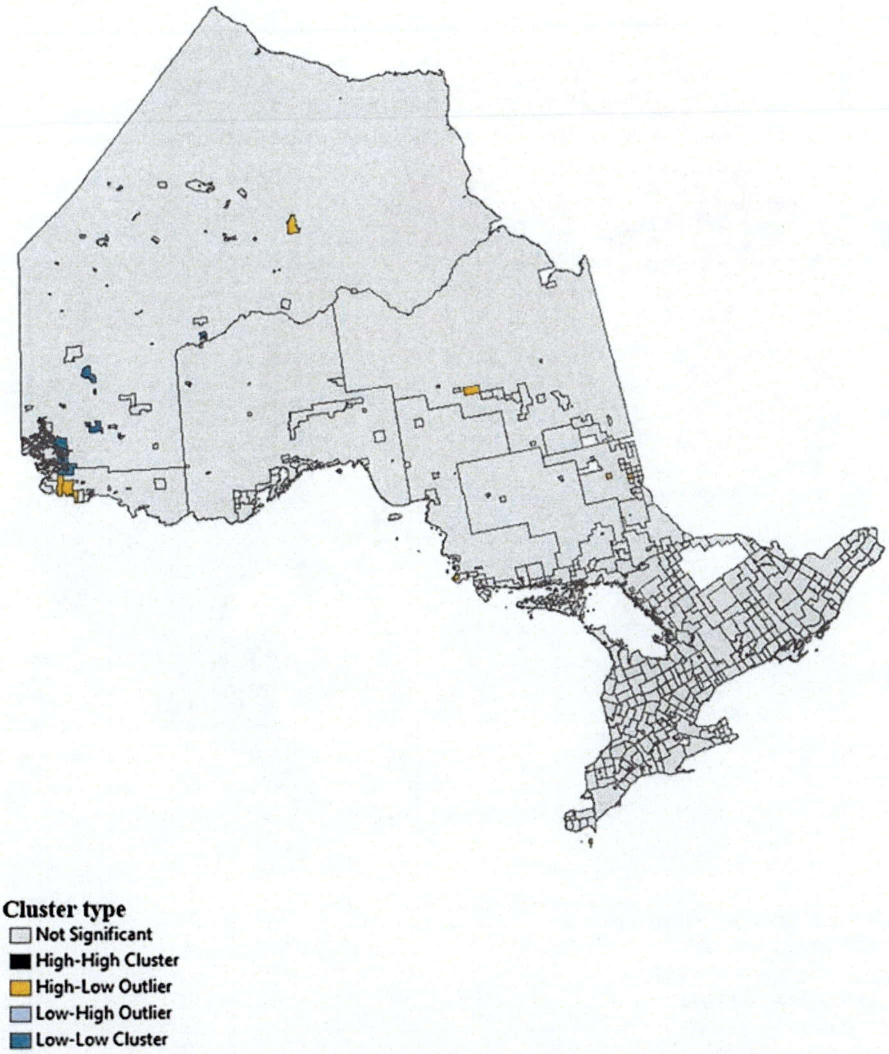

Fig. 5.7 Map of clusters—less than 5 h of caregiving per week

Model 2 (cultural background)—Does ethnic or cultural background influence the level of informal caregiving to seniors?

Table 5.10 details the R-squared results and interpretation for each model. The low values for R-squared indicate that every model was weak and poorly predicted the relationship between ethno-cultural background and rates of informal caregiving.

Model 3 (Positive Care Law)—Is there a relationship between the provision of care and the need for care in communities? Does the Positive Care Law apply to eldercare in Ontario?

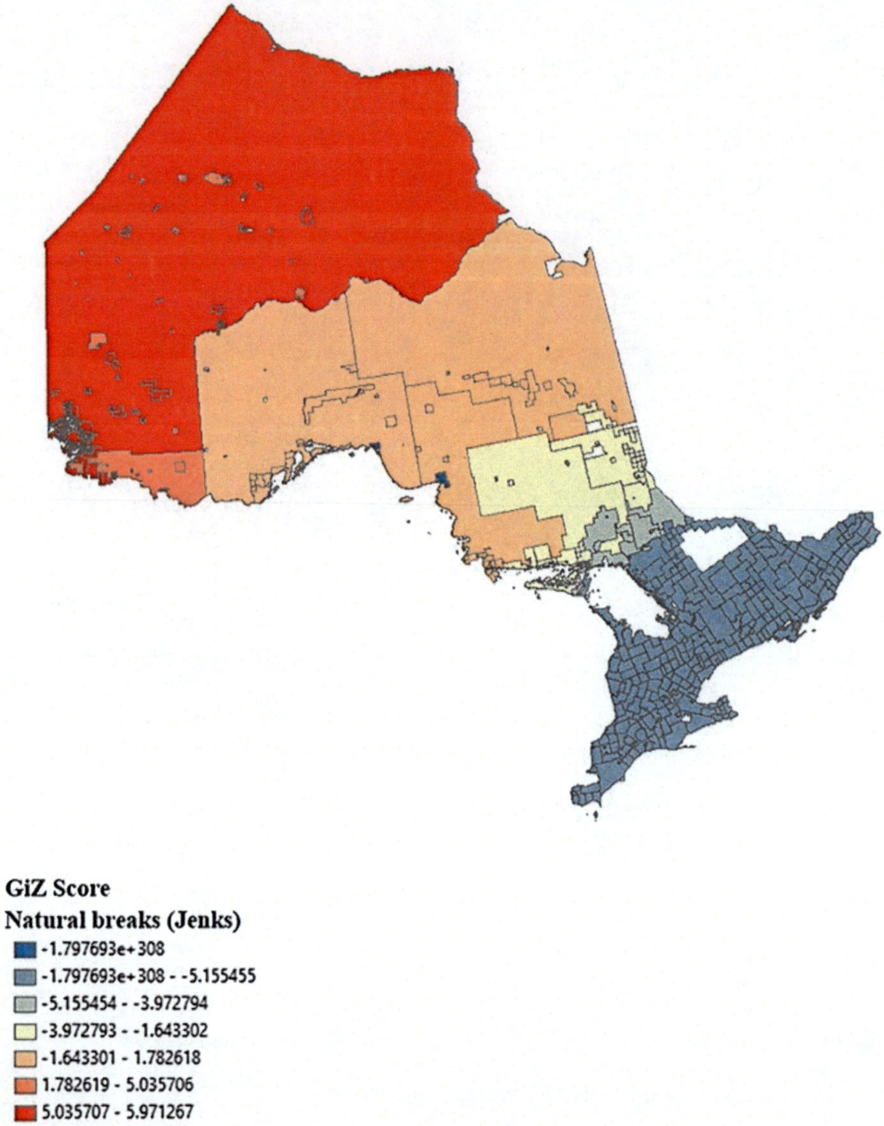

Fig. 5.8 Map of hotspots—5–9 h of caregiving per week

Table 5.11 details the R-squared results and interpretation for this model. The low R-squared values show that the model is a poor predictor of the relationship between care need and rates of caregiving.

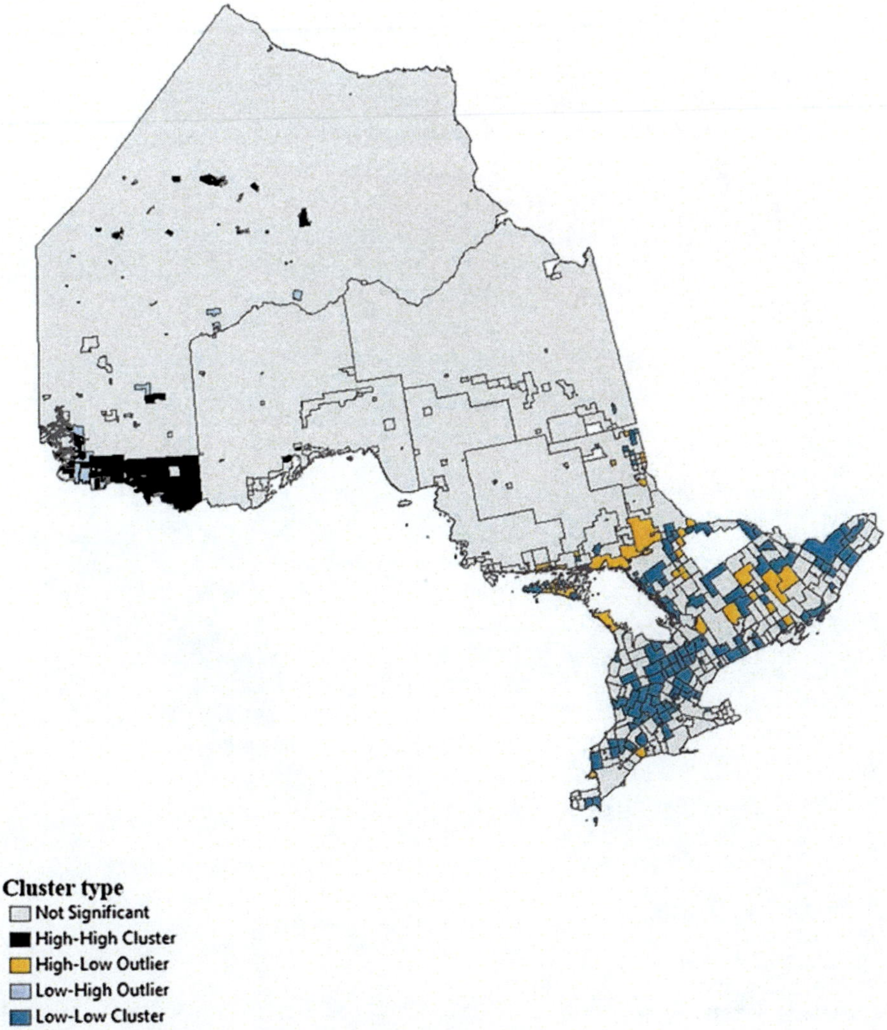

Fig. 5.9 Map of clusters—5–9 h of caregiving per week

5.5.2 *Discussion*

Care Need and Rates of Caregiving

With 1.8M Ontarians in 2006 reporting that they were engaged in caregiving, it is a very prevalent activity. Caregivers are more prevalent than Ontario seniors themselves; there were 1,649,180 Ontario seniors in 2006 (Statistics Canada 2010). This indicates that some seniors may be receiving informal care from multiple persons. The summary statistics showed that numbers of people involved in caregiving fell as the hours of caregiving rose, but rose at the highest level of 20 or more hours per

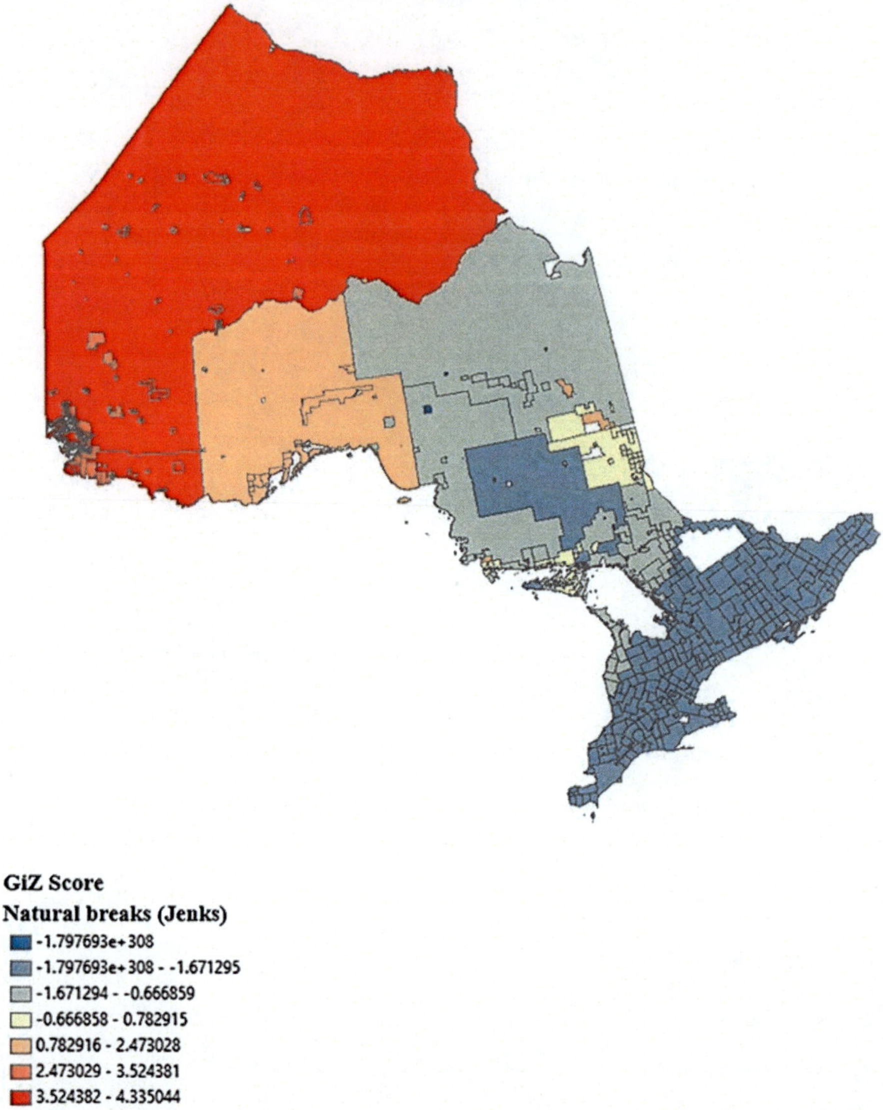

Fig. 5.10 Map of hotspots—10–19 h of caregiving per week

week. One explanation for this change may be that services are not available in some areas of high care need, so family and community members engage in more caregiving. In contrast, less than 5 h per week may be manageable for many people, but 5–19 h is not, so when care needs rise to that level, people seek out and use services where they are available.

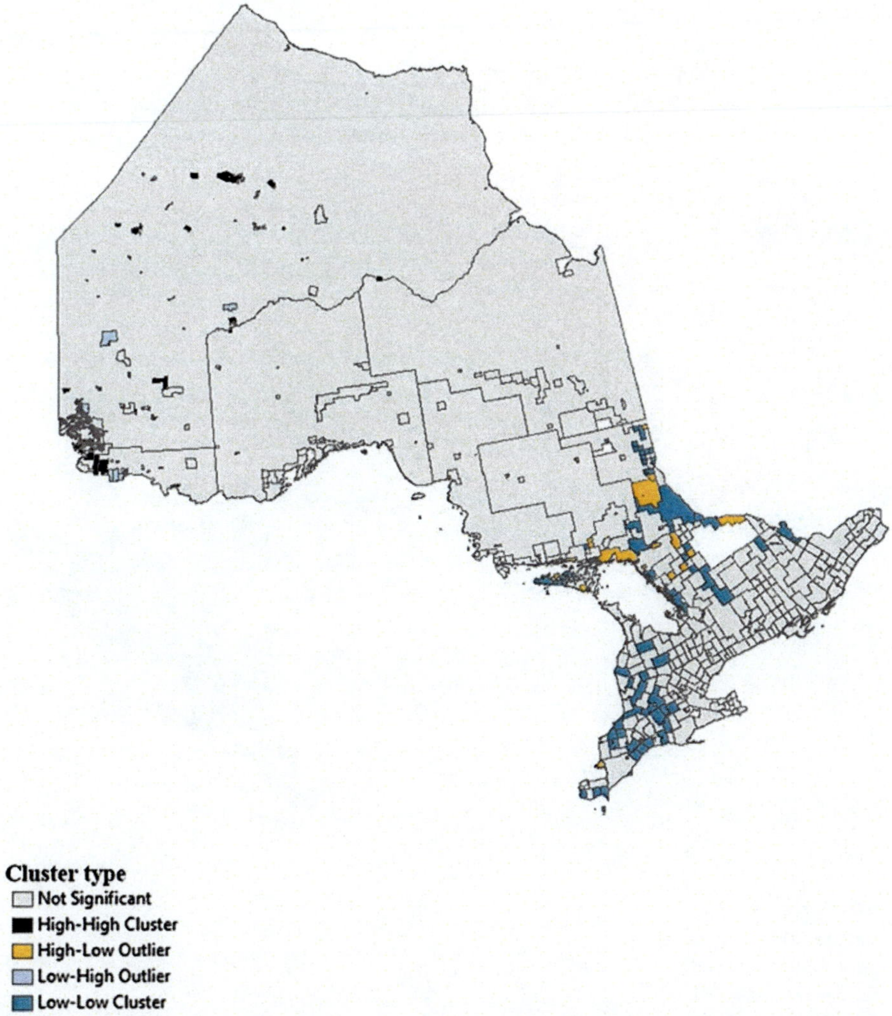

Fig. 5.11 Map of clusters—10–19 h of caregiving per week

Spatial Pattern

The results of Global and Local Moran's *I* and the Getis-Ord Gi* analyses clearly show that there is a spatial pattern to the caregiving levels across the province. Roughly, there is a North-South divide, with more caregiving in the North. There is also an urban-rural-remote division, with caregiving levels rising with distance from urban centers, reaching the very highest levels in the most remote areas. The clusters of high rates of caregiving are also mostly found in the North. The rural areas of Southern Ontario generally have clusters of low rates of caregiving, with a few high outliers, and overall are "cool spots." Where the High-High clusters, High-Low outliers, and hotspots are found, often so are First Nations communities. This is

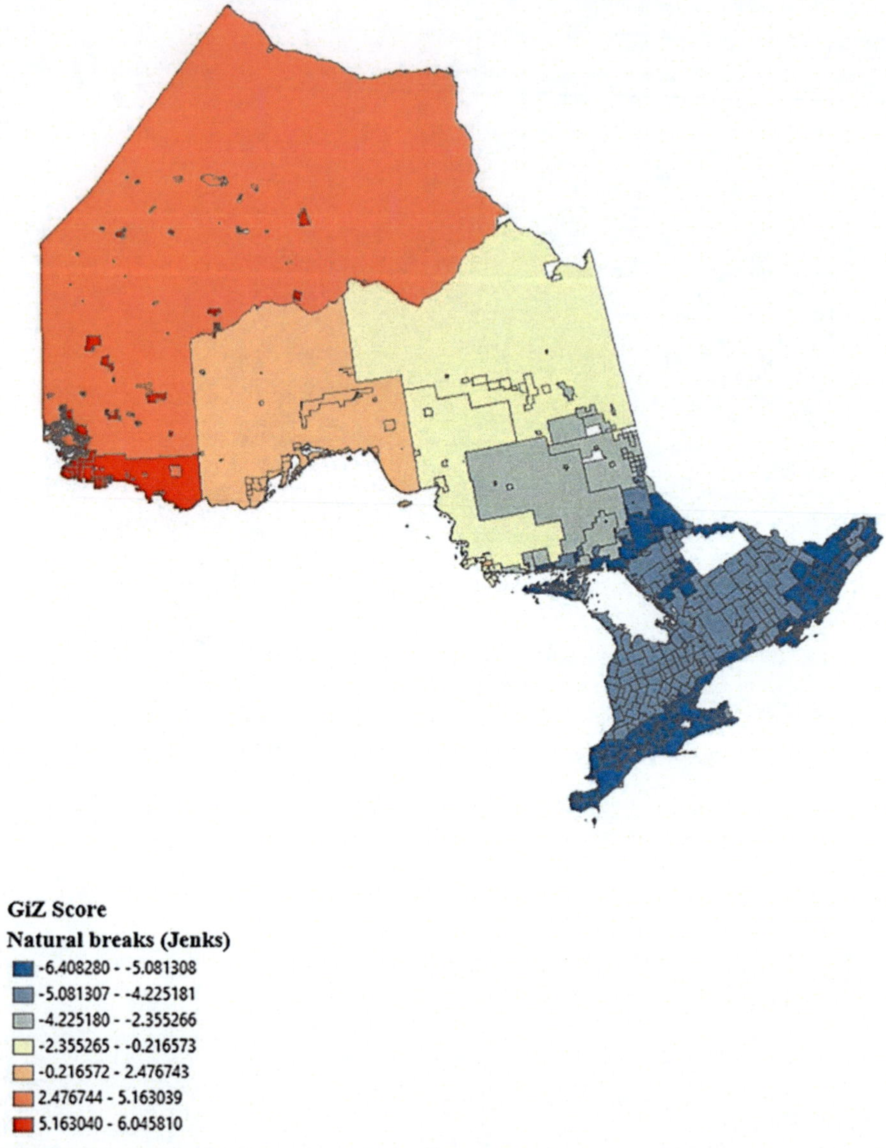

Fig. 5.12 Map of hotspots—20 h or more of caregiving per week

particularly true at the higher levels of caregiving. This spatial patterning may be reflective of care need, availability of formal care services, propensity of communities toward caregiving, or a combination of these and other factors.

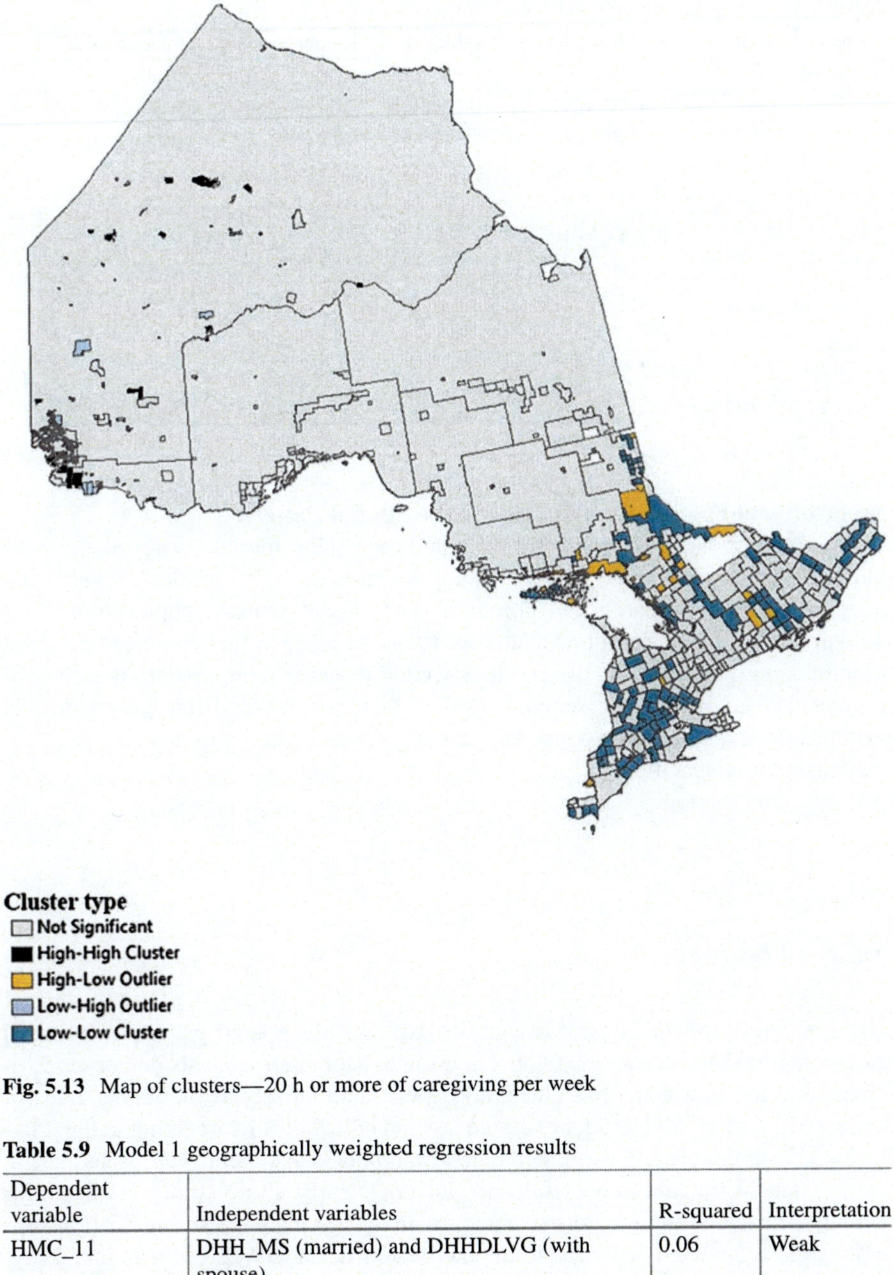

Fig. 5.13 Map of clusters—20 h or more of caregiving per week

Table 5.9 Model 1 geographically weighted regression results

Dependent variable	Independent variables	R-squared	Interpretation
HMC_11	DHH_MS (married) and DHHDLVG (with spouse)	0.06	Weak
HMC_11	DHH_MS (not married) and DHHDLVG (alone)	0.25	Weak

Table 5.10 Model 2 geographically weighted regression results

Dependent variable	Independent variable	R-squared	Interpretation
Less than 5	INDIV	0.03	Weak
	FAMIL	0.09	Weak
5–9	INDIV	0.18	Weak
	FAMIL	0.19	Weak
10–19	INDIV	0.06	Weak
	FAMIL	0.07	Weak
20 or more	INDIV	0.07	Weak
	FAMIL	0.11	Weak

Table 5.11 Model 3 geographically weighted regression results

Dependent variable	Independent variable	R-squared	Interpretation
HMC_11	ADLF6R	0.30	Weak

Sociocultural Factors Contributing to the Spatial Pattern

The poor fit of the geographically weighted regression models is not likely to be indicative of the true relationships among the variables. The results are very contrary to previous research and common sense about human relationships. It is unlikely that spouses do not provide care to one another or that care need does not prompt caregiving. Indeed, the previously cited research demonstrates that there is a great deal of informal caregiving in Canadian society, and the most plausible explanation is that it is a response to care need. A more likely explanation is that the models are mis-specified.

5.6 Conclusion

5.6.1 Findings

Using a variety of methods, this analysis explored the spatial pattern of informal caregiving to Ontario seniors. Global and Local Moran's I analysis on census variables revealed that there is a clustered pattern, particularly at the higher rates of caregiving. Getis-Ord Gi* identified the hot and coldspots of caregiving in the provinces. The geographically weighted regression models using both census and health survey data were less successful and did not clearly show strong relationships among the sociocultural factors selected for analysis. Given the strong relationships among these factors usually found in other research, the GWR results in this analysis may be attributed to model misspecification.

5.6.2 Limitations

Census data is static, point-in-time data and therefore does not address the dynamic processes of informal caregiving. Caregiving resources are likely to vary on a weekly if not daily basis as caregivers' other responsibilities and commitments ebb and flow. Care-receivers' needs are also variable, ranging from help with grocery shopping to intensive support during recovery from surgery. Because it relies heavily on census data, this analysis provides no insight into these dynamics.

Though 1.8 million Ontarians reported that they were engaged in informal caregiving on the 2006 census, it is possible that some caregivers did not, because they do not identify themselves as such. They consider caregiving simply a part of normal family responsibilities and not a separate category of activity, especially if the time and effort involved are low. By one estimate, 75% of family caregivers will not identify themselves as such until it becomes an economic and/or psychological burden (Bruhn and Rebach 2014, p. 7). This may be especially true of spousal caregivers. Even the language used to describe caregiving activities can influence caregivers' self-identification (Amaro 2015). Were the "hidden" level of informal caregiving included in this analysis, the results might be very different.

The blunt method of categorizing "individualistic" and "familistic" cultures by language most often spoken at home conceals the diversity among cultural groups and their attitudes. For example, in many predominantly First Nations communities, English was reported to be the language most often spoken at home, and so they were counted in the "individualistic" group. This is probably not at all reflective of their actual attitudes toward caregiving, given the high rates of caregiving in those communities generally, as revealed by the hotspot and Local Moran's *I* analyses. Additionally, the "individualistic" group dominates every census subdivision; it constitutes 80% of 495 of 527 CSDs, and there are only ten CSDs where it is less than 50%.

For such a large study area, the health survey data had few respondents meeting the criteria of being seniors and requiring ADL help. The data itself is highly appropriate for this analysis, as it directly addresses the topics of needing care, whether it was received and from which sources. Thus, the implausible results must be attributed to the methods applied to the survey data. This analysis would have benefitted from conducting ordinary least squares regression to develop properly specified models before proceeding to geographically weighted regression.

This analysis does not adequately address a very important catalyst to caregiving: need. The health survey data is insufficient to determine the level of care need among seniors across the province. If, as the Positive Care Law suggests, rates of informal caregiving rise as care needs rise, and care needs rise as health declines, then the maps of hotspots and clusters of caregiving may be more appropriately read as maps of senior ill-health than any other characteristic. Coldspots with lower rates of caregiving might portray areas of good health among seniors or unmet care need.

Other significant factors contributing to levels of caregiving were not part of this analysis. These include the availability of formal care services, both private and public, family dynamics, geographic proximity between caregivers and care-receivers, co-residence, and the many characteristics of caregivers themselves such as age, gender, marital status, labor force participation, and income. Statistics Canada has conducted many national studies that produced global results on these aspects of caregiving, but this analysis did not examine geographic variation among them. Finally, this analysis suffers from the scale effect of the modifiable areal unit problem. Census subdivisions involve a high level of aggregation. In a study area as large and diverse as Ontario, there are sure to be spatial processes at work at lower scales that might be significant to an understanding of caregiving behavior.

5.6.3 Future Research

Further analysis of this topic should focus on smaller study areas and smaller units of observation, such as census metropolitan areas at the census tract level. This would correct the modifiable areal unit problem and improve the analysis of spatial patterning and relationships among variables. Such an approach would be especially useful in exploring the reasons for the urban-rural-remote, North-South divide found in this study.

Ideally, both care need indicators, such as seniors' health levels, and formal care service level indicators could be integrated into caregiving analysis in order to better understand the relationships among these key contributing factors. Further analysis should also involve a more nuanced way of accounting for ethno-cultural attitudes toward caregiving. Blending several years' worth of pertinent healthy survey data might result in a more robust dataset better suited to spatial analysis.

Finally, given the complexity of the topic and the numerous factors that contribute to rates of caregiving, a better approach would involve beginning with exploratory regression analysis with many more variables, possibly dozens. This would enable the development of properly specified global models, which would assist in the correct application of other spatial analysis methods such as geographically weighted regression, and a better understanding of the data and topic generally.

Appendices

Appendix A: Census and Canadian Community Health Survey Questions

2006 census questions

Question 33c

Last week, how many hours did this person spend doing the following activities:

(c) Providing **unpaid** care or assistance to one or more seniors?

Some examples include providing personal care to a senior family member; visiting seniors; talking to them on the telephone; helping them with shopping, with banking, or with taking medication; etc.

- None
- Less than 5 h
- 5–9 h
- 10–19 h
- 20 h or more

Question 15 (a)

What language does this person speak **most often** at home?

2009 Canadian Community Health Survey questions

ADLF6R—Activities of daily living

Because of any physical condition or mental condition or health problem, do you need the help of another person:

- With preparing meals?
- With getting to appointments and running errands such as shopping for groceries?
- With doing everyday housework?
- With personal care such as washing, dressing, eating, or taking medication?
- With moving about inside the house?
- With looking after your personal finances such as making bank transactions or paying bills?

HMC_11—Receipt of nongovernment home care

Have you received any other home care services in the past 12 months, with the cost not covered by government (e.g., care provided by a private agency or by a spouse or friends)?

DHH_MS—Marital status

What is [respondent name]'s marital status? Is [he/she]:,

1 … married?
2 … living common law?
3 … widowed?
4 … separated?
5 … divorced?
6 … single, never married?

DHHDLVG—Living arrangement

This variable was derived from a series of questions regarding the respondent's living arrangements and relationships to others with whom they live.

Appendix B: Language Groups Used to Create INDIV and FAMIL Variables

INDIV variable

English	Flemish	Swedish
French	Frisian	Finnish
Danish	German	
Dutch	Norwegian	

FAMIL variable

Algonquin	Yiddish	Kurdish
Atikamekw	Bosnian	Panjabi (Punjabi)
Blackfoot	Bulgarian	Pashto
Carrier	Croatian	Persian (Farsi)
Chilcotin	Czech	Sindhi
Chipewyan	Macedonian	Sinhala (Sinhalese)
Cree	Polish	Urdu
Siouan languages (Dakota/Sioux)	Russian	Malayalam
Dene	Serbian	Tamil
Dogrib	Serbo-Croatian	Telugu
Gitksan	Slovak	Japanese
Inuinnaqtun	Slovenian	Korean
Inuktitut, n.i.e.	Ukrainian	Cantonese
Kutchin-Gwich'in (Loucheux)	Latvian	Chinese, n.o.s.
Malecite	Lithuanian	Mandarin
Mi'kmaq	Estonian	Taiwanese
Mohawk	Hungarian	Lao
Montagnais-Naskapi	Greek	Khmer (Cambodian)
Nisga'a	Armenian	Vietnamese
North Slave (Hare)	Turkish	Bisayan languages
Ojibway	Amharic	Ilocano
Oji-Cree	Arabic	Malay
Shuswap	Hebrew	Tagalog (Pilipino, Filipino)
South Slave	Maltese	Akan (Twi)
Tlingit	Somali	Swahili
Italian	Tigrigna	Creoles
Portuguese	Bengali	Other languages
Romanian	Gujarati	
Spanish	Hindi	

References

AARP (2013) You take care of Mom, but who will take care of you? Retrieved on 15 April 2014 from http://www.aarp.org/home-family/caregiving/info-08-2013/the-aging-of-the-baby-boom-and-the-growing-care-gap-AARP-ppi-ltc.html

Administration on Aging (AoA) (2015). National survey of older Americans Act participants. Retrieved on 17 August 2015 from https://aoasurvey.org/source/priorresults.asp

Aging in Place Technology Watch (2015) Retrieved on 15 August 2015 from http://www.ageinplacetech.com/

Amaro LM (2015) Convincing family caregivers to participate in research: "Does that count as caregiving?". J Appl Commun Res 43(2):267–272

Anonymous. (2010) Healthy profits from caregiving: growth driven by demographics and need to control costs. Shareowner 23(3):17–22

Anselin L (1995) Local indicators of spatial association-LISA. Geogr Anal 27(2):93–115

Briggs R (2010) Local measures of spatial autocorrelation. Retrieved 19 August 2015 from http://www.utdallas.edu/~briggs/henan/11SA.ppt

Bruhn JG, Rebach HM (2014) The sociology of caregiving. Springer, Dordrecht

Canadian Institute for Health Information (CIHI) (2010) Supporting informal caregivers: the heart of home care. Retrieved on 16 July 2015 from https://secure.cihi.ca/free_products/Caregiver_Distress_AIB_2010_EN.pdf

Center for Technology and Aging (2014) The new era of connected aging: a framework for understanding technologies that support older adults aging in place. Retrieved on 10 August 2015 from http://www.techandaging.org/ConnectedAgingFramework.pdf

Chappell NL (1985) Social support and receipt of home care services. Gerontologist 25(1):47–54

Chappell NL, Funk L (2012) Filial responsibility: does it matter for care-giving behaviours? Ageing Soc 32:112–1146

Choi NG (1994) Patterns and determinants of social services utilization: comparison of the childless elderly and elderly parents living with or apart from their children. Gerontologist 34(3):353–362

Colombo F et al (2011) Help wanted? Providing and paying for long-term care, OECD health policy studies. OECD Publishing, Paris

Donovan R, Williams AM (2015) Care-giving as a Canadian-Vietnamese tradition: 'It's like eating, you just do it'. Health Soc Care Commun 23(1):79–87

Encyclopedia Britannica (2015) Ontario. Retrieved on 31 May 2015 from http://www.britannica.com/EBchecked/topic/429340/Ontario

ESRI (2015) How spatial autocorrelation (Global Moran's *I*) works. Retrieved on 20 June 2015 from http://resources.arcgis.com/en/help/main/10.1/index.html#/How_Spatial_Autocorrelation_Global_Moran_s_I_works/005p0000000t000000/

Expert Group on Home and Community Care (2015) Bringing care home: report of the Expert Group on Home and Community Care. Retrieved on 10 August 2015 from http://health.gov.on.ca/en/public/programs/ccac/docs/hcc_report.pdf

Foley R (2002) Assessing the applicability of GIS in a health and social care setting: planning services for informal carers in East Sussex, England. Soc Sci Med 55:79–96

Foley R (2008) Geographies of informal care in Ireland, 2002–2006. Ir Geogr 41:261–278

Foley R, Charlton M, Fotheringham AS (2009) GIS in health and social care planning. In: Bavaud F, Mager C (eds) Handbook of theoretical and quantitative geography. FGSE, University of Lausanne, Lausanne, pp 73–115

Fowler CS (2011) Geographically weighted regression: CSDE statistics workshop, February 1st 2011. Center for Studies in Demography and Ecology, University of Washington. Retrieved on 10 November 2013 from http://csde.washington.edu/services/gis/workshops/Resources/GWR_Presentation.pdf

Freedonia Group (2015) Continuing care retirement communities: United States. Retrieved on 23 March 2015 http://www.freedoniagroup.com/freedonia-focus/FF40058/elder-care-services.htm

Gaymu J, Busque M-A, Légaré J, Décarie Y, Vézina S, Keefe J (2010) What will the family composition of older persons be like tomorrow? A comparison of Canada and France. Can J Aging 29(1):57–71

Getis A, Ord JK (1992) The analysis of spatial association by use of distance statistics. Geogr Anal 24(3):189–206

Global Action on Aging (2015) The 5 year review of the Madrid international plan of action on ageing. Retrieved on 17 August 2015 from http://www.globalaging.org/agingwatch/events/CSD/mipaa+5.htm

Goodchild MF (1986) Spatial auto-correlation, CATMOG 47. GeoBooks, Norwich

Government of Canada (2013) Information for caregivers: federal, provincial and territorial resources. Retrieved on 6 January 2015 from http://www.seniors.gc.ca/eng/sb/caregivers/index.shtml

Government of Canada (2014) Canadian employers for caregivers plan. Retrieved on 6 January 2015 from http://www.esdc.gc.ca/eng/seniors/cecp/index.shtml

Government of Manitoba (2014) Primary caregiver tax credit. Retrieved on 6 January 2015 from http://www.gov.mb.ca/finance/tao/caregiver.html

Government of Nova Scotia (2014) Caregiver benefit. Retrieved on 6 January 2015 from http://novascotia.ca/DHW/CCS/FactSheets/Caregiver-Benefit.pdf

Grant K, Church E (2015) No place like home? Investigating Ontario's home-care shortcomings. The Globe and Mail

Griffith DA (1987) Spatial auto-correlation: a primer. Association of American Geographers, Washington DC

Health Canada (2003) Respite for family caregivers: an environmental scan of publicly-funded Programs in Canada. Retrieved on 6 January 2015 from http://www.hc-sc.gc.ca/hcs-sss/alt_formats/hpb-dgps/pdf/pubs/2003-respite-releve/2003-respite-releve-eng.pdf

Health Council of Canada (2012) Seniors in need, caregivers in distress: what are the home care priorities for seniors in Canada? Retrieved on 6 August 2015 from http://www.alzheimer.ca/on/~/media/Files/on/Media%20Releases/2012/April%202012/HCC_HomeCare_2d.pdf

Hirshorn B, Stewart JE (2003) Geographic Information Systems in Community-Based Gerontological Research and Practice. J Appl Gerontol 22(2):134–151

Hollander MJ, Guiping L, Chappell N (2009) Who cares and how much: the imputed economic contribution to the Canadian healthcare system of middle-aged and older unpaid caregivers providing care to the elderly. Healthcare Q 12(2):42–49

Johnson RW (2008) Choosing between paid elder care and unpaid help from adult children: the role of relative prices in the care decision. In: Szinovacz ME, Davey A (eds) Caregiving contexts: cultural, familial and societal implications. Springer, New York, pp 35–69

Kaida L, Mayer M, Park SY (2009) Cultural preferences and economic constraints: the living arrangements of elderly Canadians. Can J Aging 28(4):303–313

Kitchen P, Williams A, Pong RW (2011) Socio-spatial patterns of home care use in Ontario, Canada: a case study. Health Place 17:195–206

LaPierre TA, Keating N (2013) Characteristics and contributions of non-kin carers of older people: a closer look at friends and neighbours. Ageing Soc 33:1442–1468

Lowenstein A, Katz R, Gur-Yaish N (2008) Cross-national variations in elder care: antecedents and outcomes. In: Szinovacz ME, Davey A (eds) Caregiving contexts: cultural, familial and societal implications. Springer, New York, pp 93–112

Mathur A, Moschis GP (1999) Exploring the intergenerational caregiver market: a study of family care providers for the elderly. J Mark Theory Pract 7(3):76–86

Mitchell BA (2001) Ethnocultural reproduction and attitudes towards cohabiting relationships. Can Rev Sociol Anthropol 38(4):391–413

Neiman FD (2007) Local spatial autocorrelation. Retrieved on 18 August 2015 from http://people.virginia.edu/~fn9r/anth589b/LISALectureSlides.pdf

Ontario Human Rights Commission (2015) The duty to accommodate. Retrieved on 13 July 2015 from http://www.ohrc.on.ca/en/policy-and-guidelines-discrimination-because-family-status/vi-duty-accommodate

Ontario Ministry of Health (MOH) (2015) Home First – Putting Patients at the Centre of their Health Care, Retrieved on 6 August 2015 from http://www.health.gov.on.ca/en/pro/programs/ecfa/action/community/com_homefirst.aspx

Ontario Ministry of Labour (MOL) (2015) Family caregiver leave. Retrieved on 13 July 2015 from http://www.labour.gov.on.ca/english/es/pubs/guide/caregiver.php

Palo Stoller E, Schroeder Miklowski C (2008) Spouses caring for spouses: untangling the influences of relationship and gender. In: Szinovacz ME, Davey A (eds) Caregiving contexts: cultural, familial and societal implications. Springer, New York, pp 115–131

Peek MK, Coward RT, Peek CW, Lee GR (2008) Are expectations of care related to the receipt of care? An analysis of parent care among disabled elders. J Gerontol 53B(3):S127–S136

Pong R (2008) Informal care: some findings from research studies. Slides from the Change Foundation Event, Informal Care. Toronto. June 2008. Retrieved on 10 May 2014 from http://www.changefoundation.ca/library/informal-care-some-findings-from-research-studies

Rosenshein L (2011) Local Moran's I: computing (Online forum comment). Retrieved on 30 August 2015 from https://geonet.esri.com/thread/19745

Schwanen T, Hardill I, Lucas S (2012) Spatialities of ageing: the co-construction and co-evolution of old age and space. Geoforum 43:1291–1295

Scott L (2008) Moran's I vs. Getis-Ord General G (Online forum comment). Retrieved on 30 August 2015 from http://forums.esri.com/Thread.asp?c=93&f=983&t=270745#835279

Shaw M, Dorling D (2004) Who cares in England and Wales? The Positive Care Law: cross-sectional study. Br J Gen Pract 54:899–903

Shoval N, Wahl HW, Auslander G, Isaacson M, Oswald F, Edry T, Landau R, Heinik J (2010) Use of the global positioning system to measure the out-of-home mobility of older adults with differing cognitive functioning. Ageing Soc 31:849–869

Statistics Canada (2003) Unpaid informal caregiving. Catalogue No. 11-008

Statistics Canada (2008) Projecting future availability of the informal support network of the elderly population and assessing its impact on home care services. Catalogue no. 91F0015M – No. 009

Statistics Canada (2009a) Hours spent providing unpaid care or assistance to seniors. Retrieved on 30 April 2014 from http://www12.statcan.ca/census-recensement/2006/ref/dict/pop045-eng.cfm

Statistics Canada (2009b) Canada's changing labour force, 2006 Census: the provinces and territories. http://www12.statcan.gc.ca/census-recensement/2006/as-sa/97-559/p32-eng.cfm. Accessed 15 May 2014

Statistics Canada (2010) Age and Sex, 2006 counts for both sexes, for Canada, provinces and territories – 100% data. Retrieved on 3 September 2015 from http://www12.statcan.gc.ca/census-recensement/2006/dp-pd/hlt/97-551/pages/page.cfm?Lang=E&Geo=PR&Code=01&Table=1&Data=Count&Sex=1&StartRec=1&Sort=2&Display=Page

Statistics Canada (2011) Caring for a parent who lives far away: the consequences. Publication 11-008-X. Retrieved on 21 March 2014 from http://www.statcan.gc.ca/pub/11-008-x/2010001/article/11072-eng.htm

Statistics Canada (2012a) Classification of unpaid care. Retrieved on 30 April 2014 from http://www.statcan.gc.ca/concepts/definitions/unpaidclass-impayeclass02-eng.htm

Statistics Canada (2012b) Informal caregiving for seniors. Catalogue no. 82-003-X

Statistics Canada (2012c) Living arrangements of seniors. Catalogue no. 98-312-X20110003

Statistics Canada (2012d) Portrait of caregivers, 2012. Catalogue no. 89-652 – No. 001

Statistics Canada (2013) The general social survey: an overview. Retrieved on 6 January 2015 from http://www.statcan.gc.ca/pub/89f0115x/89f0115x2013001-eng.htm#a4

Statistics Canada (2014a) Receiving care at home. Catalogue no. 89-652-X – No. 002

Statistics Canada (2014b) Canadians with unmet home care needs. Catalogue no. 75-006-X

Statistics Canada (n.d.) Canadian Community Health Survey (CCHS) – healthy aging. Retrieved on 6 January 2015 from http://www23.statcan.gc.ca/imdb/pIX.pl?Function=showStaticArchiveHTML&a=1&fl=http://www23.statcan.gc.ca/imdb-bmdi/instrument/5146_Q1_V2-eng.htm&Item_Id=53430#a27

The British Columbia Law Institute and the Canadian Centre for Elder Law (2010) Care/work: law reform to support family caregivers to balance paid work and unpaid caregiving. Author, Vancouver

The Conference Board of Canada (2013) Future care for Canadian seniors – why it matters. Retrieved on 6 January 2015 from http://www.conferenceboard.ca/e-library/abstract.aspx?did=5753

United Nations Department of Economics and Social Affairs (2008) Principles and recommendations for population and housing censuses: revision 2. New York. Retrieved on 12 January 2015 from http://unstats.un.org/unsd/demographic/sources/census/docs/P&R_Rev2.pdf

United Nations Population Fund (UNFPA) and HelpAge International (2012) Ageing in the 21st century: a celebration and a challenge. Retrieved on 15 August 2015 from http://www.unfpa.org/publications/ageing-twenty-first-century

United Nations Second World Assembly on Ageing (SWAA) (2002) Madrid international plan of action on ageing. Retrieved on 17 August 2015 from http://www.un.org/en/events/pastevents/pdfs/Madrid_plan.pdf

Van Loon J (2008) Geographically Weighted Regression: presentation for GEOG 570. Department of Geography, University of British Columbia. Retrieved on 10 November 2013 from http://www.geog.ubc.ca/courses/geog570/presentations/GWR_FEB_26_08_JvL.pdf

Young H, Grundy E, Kalogirou S (2005) Who cares? Geographic variation in unpaid caregiving in England and Wales: evidence from the 2001 Census. Popul Trends 120:23–34

Young-Hoon KN (2006) Relationship between home care services and informal caregivers. Lakehead University, Thunder Bay

Chapter 6
A Spatial Statistical Understanding of Inadequate Dwellings in the City of Toronto

6.1 Introduction

6.1.1 Context

This study is inspired by my personal lived experiences with poor dwelling conditions in both subsidized and non-subsidized housing and the announcement of the federal government's plans for housing repairs through a National Housing Strategy. Poor dwelling conditions have been a factor and an output in the study area of the City of Toronto, contributing to a polarization in wealth and quality of life (Hulchanski 2010). Housing is a key contributor to quality of life and is a building block to other areas such as personal relationships, education, the workforce, and community engagement (CMHC 2014). Tenants with poor dwelling conditions are more likely to be negatively impacted by homes in need of repairs that are inadequate (United Way 2011). Under dwelling characteristics—the physical features of residences—dwelling conditions are an indicator of housing adequacy and are classified into groups requiring regular maintenance, minor repairs, or major repairs, with the latter considered inadequate by organizations interested in housing (Statistics Canada 2017).

These issues have prompted the federal government to create a National Housing Strategy which plans to prioritize vulnerable Canadians with investments and programs to tackle: homelessness, housing affordability, suitability, and adequacy (Government of Canada 2017). The understanding of poor dwelling conditions within a National Housing Strategy or any other form of intervention can benefit from statistical and spatial analysis to identify what relationships and circumstances are present. Knowing where and what has a significant relationship with poor dwelling conditions can inform decision-makers in targeting adequate investments and programs.

E. Vaz, *Regional and Urban Change and Geographical Information Systems and Science*, Advances in Geographic Information Science,
https://doi.org/10.1007/978-3-031-24731-6_6

6.1.2 Research Questions and Objectives

Achieving the statistical and spatial knowledge on dwelling conditions can be succeeded by answering the following research questions:

1. Which spatially explicit census variables suggest a significant relationship with inadequate dwelling conditions and how it has changed in the census years of the 1990s, 2000s, and 2010s by census tract?
2. Focusing on the 2016 census, what is the statistical and spatial relationship between inadequate dwelling conditions and the percentage of subsidized and non-subsidized tenant dwellings by census tract?

Answering the previously mentioned research questions can be supplemented by an undertaking of the following research objectives:

1. Using the 2016 census at the census tract level, collect variables based on the literature review that could have a relationship with the dependent variable of inadequate dwelling conditions while also ensuring the same variables are available in the census years of 1996, 2001, 2006, and 2011.
2. Improve the list of census variables by using a hierarchal cluster analysis and stepwise regression model to ensure independence and significant relationships.
3. Conduct a number of multivariate regression models to achieve the best explanatory model for dwelling conditions and an acceptable R^2.
4. Once the ideal census variables are selected from the 2016 census explanatory model, the process will be repeated with identical census variables, for the previous census years of interest.
5. Statistical outputs such as the coefficient of determination (R-squared) and inferred regression residuals for maps will be recorded to measure the differences over the census years of the past 20 years.
6. Solely working with the 2016 census again, the regression residuals of that year will be used in a bivariate analysis with the census variables on the percentage of tenant subsidized dwellings and a derived percentage of tenant non-subsidized dwellings.
7. A second bivariate analysis will be conducted between the two tenant subsidized statuses and the inadequate dwellings variable, to compare bivariate results.

Completing these research objectives will not only answer the research questions in this major research paper but will also form the methodology and produce results for discussion. Predictors or census variables will be understood and grouped into dwelling or household/resident indicators that have a relationship with dwelling conditions. This understanding will also inform the recent 2016 census and the relationship between subsidized tenant status and inadequate dwelling conditions.

6.2 Literature Review

6.2.1 History and Background

Dwelling conditions became a concern several years after the Industrial Revolution resulting in crowded dwellings and a necessary focus on public health (Filali 2012). The relationship between human health and dwelling condition can also stem from the nineteenth-century research of John Snow and has evolved to be more conceptual in understanding housing and social determinants (Baker et al. 2016a). The link between disease and poor dwelling conditions has also been known since the nineteenth century in the work of Edwin Chadwick in Great Britain on sanitary conditions and the laboring population (Ellen and Glied 2016).

Canada in the twentieth century has been able to provide adequate housing to the majority of Canadian workers, but housing inadequacy has been an issue that has not improved for marginalized communities (Purdy 2003). The 25 years before 2003 have seen a decline in economic and social well-being for the Canadian working class, affecting housing opportunities such as adequate dwelling conditions, especially for low-income visible minority families (Purdy 2003). According to the 1996 Canadian census, immigrants were more likely to have core housing needs, including dwelling condition issues, than their non-immigrant counterparts (Murdie 2003). Dwelling condition has been an issue with tenants in Toronto for years, especially for immigrant and visible minority groups such as the Bengali community causing housing shame and dissatisfaction (Ghosh 2014).

According to the United Nations, everyone should have an adequate standard of living for themselves and their loved ones, as well as continuous improvement of their dwelling and living conditions, regardless of background (CESCR 1991). In regard to dwelling interiors, an adequate and habitable home should have sustainable accessible energy for cooking, heating, lighting, washing facilities, food storage, and safe drinking water (CESCR 1991). Public or subsidized housing in particular has an obligation to meet standards on housing adequacy set by the United Nations for the benefit of low-income people, as well as individual and community health (Dahmann and Dennison 2013).

Research from several social science disciplines have been concerned with dwelling conditions as an issue regarding ones quality of life (Morton et al. 2004).

Housing is beyond a form of shelter, but a crucial part of life that affects educational and employment opportunities, as well as wealth, health, and well-being (Baker et al. 2017). Satisfaction in a dwelling space is based on a level of agreement between desired and actual housing attributes, while dissatisfaction is based on the discrepancy between aspirations and reality (Gibler et al. 2014).

Housing adequacy is a crucial part of understanding housing quality and differs from affordability and availability since it depends on government regulations concerned with the physical characteristics of dwellings (Morton et al. 2004). Multiple criteria can be used to measure quality, which can include structural quality, caretaking or the state of maintenance, and service quality which is based on the conveniences, facilities, and equipment the dwelling offers (Djebarni and Al-Abed 1998). Adequacy can also be understood by the configurations, dimensions, or rooms of a household, and if it is ideal for the residents (Campagna 2016). Without repair, the persistent breakdown of dwellings and their conditions in subsidized low-income housing creates additional cost for the organizations and governmental bodies in charge of tenant and dwelling well-being (Govender et al. 2011). Recently, public housing has been victimized by new developments and events such as the Olympics, which have often resulted in displacement and issues of housing adequacy, removing both the choice of where to live and what conditions to live in (Suzuki et al. 2018).

6.2.2 Home Characteristics and Dwelling Condition Impact

The quality of life of residents can be understood by dwelling quality characteristics which include dwelling performance requirements, dwellers' characteristics, and space characteristics (Tibesigwa et al. 2017). A key dimension of housing deprivation is housing inadequacy, which comprises structural deficiency and a lack of dwelling facilities (Filandri and Olagnero 2014). An important part of housing adequacy is the dwelling interior which deals with internal concerns such as lighting, heating, ventilation, dampness, plumbing, the kitchen, instability, occupancy, and building age (Djebarni et al. 1998). Interior inadequacy can be seen as a significant measure of emotional and physical well-being in the home, which takes standards of safety, size, attractiveness, and comfort into consideration (Campagna 2016). In the United Kingdom, research showed that low-quality or insecure housing may support the risk of poverty or aggravate the effects of poverty on living conditions and life opportunities (Tunstall et al. 2013).

In Beirut, Lebanon, research found that disadvantaged homes with poor dwelling conditions and multiple issues were strongly associated with residents having chronic illnesses (Habib et al. 2009). Studies suggest that dwelling quality has a positive correlation with mental health and well-being, meaning instances of mold, infestations, structural deficiencies, and housing dissatisfaction will have an undesirable impact on wellness (Evans et al. 2003). Present and long-term housing condition issues have a strong relationship with mental health and its deterioration, according to the research of Pevalin, Reeves, and others (2017). The ability to handle physical and mental problems is lowered when the condition of their dwelling is crowded, cold, noisy, and dilapidated (Howden-Chapman et al. 1996). Crowding, noise, and poor lighting can negatively affect mental health, leading to mood disorders, insomnia, and depression (Meltzer and Schwartz 2016). When homes are overcrowded, the risk of disease and its ability to spread increases, which includes physical and mental illnesses (Howden-Chapman et al. 1996). Overcrowding has been linked to poor parenting, poor martial relations, as well as physical and psychological withdrawal, and limited study space for children (Pendall et al. 2012).

Homes with extreme temperatures, whether hot or cold, are known to negatively affect children and the elderly since they spend more time at home, resulting in infectious and respiratory issues (Corman et al. 2016). Residential mobility can also be a housing issue since a high frequency of moving by a family is seen as residential instability and worsens other housing issues (Leventhal and Newman 2010). In the choice between dwelling condition and food for low-income families, housing condition is more often less prioritized leading to housing inadequacy (Corman et al. 2016).

Subsidized housing programs which result in lower rent are typically associated with lower quality and inferior energy efficiency for a dwelling and are known as filtering: older buildings becoming ideal and more affordable for low-income individuals (Grösche 2010). Older rented homes can be a problem for low-income families because of higher maintenance and energy costs, and can be slow to adapt to the needs of growing households, resulting in high costs for property owners (Pendall et al. 2012). Other research has pointed out that housing costs and its associated stress are decreased in public housing, but this does not automatically improve well-being and health, especially since housing inadequacies in aged homes may be present (Ruel et al. 2010).

In New Zealand, rented dwellings compared to owner-occupied dwellings were in worse conditions, due to a range of issues such as interior fittings, safety, moisture, insulation, and heating (Chisholm et al. 2017). On Twitter in New Zealand, housing issues became a social media campaign and discussion, with the highest percentage of tweets on structural issues focusing on tenant homes in the private rental sector (Chisholm and O'Sullivan 2017). Subsidized housing can come in various forms such as those for low-income seniors, the homeless in need of greater support, for families, vouchers to be used in private housing, mixed-income communities, and agreed tax breaks for developers (Brisson and Covert 2015). In the United States, housing satisfaction, including housing adequacy between renters of subsidized (housing vouchers) and non-subsidized homes, tends to show higher levels of satisfaction among the residents of subsidized dwellings due to offset costs (Ross et al. 2012).

Research in Atlanta found that subsidized homes are seen as a safety net as opposed to a cause of poor health by new residents and that housing improvements should be prioritized for the benefit of relocated residents (Ruel et al. 2010). Housing issues such as dwelling condition may be aggravated due to the location of both subsidized and non-subsidized homes in poorer and segregated neighborhoods of some major American cities (Talen and Koschinsky 2014). According to the economic theory, it is suggested that the organization of subsidized housing programs might lead to low upkeep leading to homes meeting minimal standards and the risk of poor dwelling conditions (Walters 2009). It should be noted that the dwelling conditions of subsidized housing, especially inadequate conditions, have a small role compared to the concentration and location of disadvantaged communities in the relationship between crime and subsidized housing (Lens 2013).

6.2.3 *Resident Characteristics and Dwelling Condition Impact*

Inadequate and unsafe dwelling conditions come in a variety of forms such as lead paint causing neurological and developmental problems in children, mold and moisture causing respiratory problems, infestations causing other diseases, and injury from falls (Meltzer et al. 2016). Research by Díaz McConnell (2017) acknowledges that in vulnerability research, older ages, racial and visible minorities, immigrants, poverty, low-income, and having English as a second language are factors linked with precarious housing. Socio-economic characteristics that affect dwelling satisfaction including marital status, sex, age, children, and income can be further affected by sociocultural traditions and housing market characteristics (Gibler et al. 2014). A household's composition and ages of its members, which can lead to the issue of crowding, can increase a dwelling condition issue and therefore depravity when it comes to the question of adequacy (Filali 2012).

Economic or occupational struggles by the adults in the household can effect child well-being and lead to inadequate or unsafe dwellings (Holupka and Newman 2011). Income is positively associated with social participation and health; with lower incomes, the elderly are more likely to suffer from poor dwelling conditions than their wealthy counterparts (Choi 1999). Those with less disposable income end up in low-income housing due to fewer choices and less income for maintenance, repairs, and some other necessities which can affect well-being (Govender et al. 2011). Persistent poor dwelling conditions may be influenced partially by inadequate financial resources and a lack of autonomy, with renters dependent on their landlords to fix dwelling problems (Pevalin et al. 2017). Poor dwelling conditions can negatively impact the choice of residents to pursue self-employment, since the space available at home (if any) would affect social activities, health, and the possibility to achieve higher earnings (Reuschke 2016). Negative consequences such as poor dwelling conditions and housing insecurity are likely to occur as rent burden increases, as well as the risk of unemployment and eviction based on a study on housing and the great recession in the United States (Colburn and Allen 2018).

Features of housing policies that are acknowledged include physical housing quality and subsidized housing, when it comes to the discussion of housing and child development (Leventhal et al. 2010). Poor amenities in the home reported by the parents were seen as harmful to their children's social adjustment at school and were associated with having psychological distress (Rollings et al. 2017). The need for poor families to afford dwellings may also result in poor child development due to moving frequently, resulting in choosing poor dwelling conditions and an increase in negative consequences (Holupka et al. 2011).

Due to retirement, the elderly usually spend their time in the home and so their well-being and the well-being of their homes become a high importance for the sake of maintaining independence (Choi 1999). For the elderly, ideal dwelling conditions allow the following household activities to be fulfilled with autonomy: climbing stairs, using sanitary and kitchen facilities, moving around the dwelling, and entering and exiting the home (Braubach and Power 2011). For older people, health and staying mobile are considered important, and any issues regarding these priorities can worsen with poor dwelling conditions, negatively impacting their autonomy (Windle et al. 2006). Older individuals are likely to reduce their outdoor hobbies for indoor hobbies such as visits, watching television, reading, and resting, resulting in a need for a comfortable home to be independent and express oneself (Braubach et al. 2011). Self-reported health status of the elderly in Wales found that homeowners had the best score, followed by those in non-subsidized rented homes, and the worst score was given by rented subsidized dwellings (Windle et al. 2006).

In Victoria, Australia, vulnerable groups like the Aboriginal Peoples are renting their homes and are unable to make structural changes to improve dwelling conditions that would eventually lead to improved resident well-being (Bedggood et al. 2017). A significant sample of New Zealand social housing tenants that were of Maori or Pacific ethnicity had low household incomes and experienced high levels of environmental tobacco smoke and crowding due to a shortage of bedrooms (Baker et al. 2016b). Individuals with mental health illnesses or psychological distress are more likely to be housed in homes with plumbing issues, dampness, and poor structural quality (Wells and Harris 2007). Negative health effects of housing have had a harder impact to vulnerable communities such as persons with disabilities and long-term tenants (Baker et al. 2017).

6.2.4 Dwelling Conditions and Interventions

In the United States, there is no widely accepted standardized system which links poor dwelling conditions and health outcomes, and dwelling conditions are assessed visually that may or may have already adversely affected resident health (Jacobs 2006). That said, developed nations and regions have pursued plans to improve housing conditions, including social housing. One of the main motivations behind subsidized housing and community development programs is to create healthier environments within dwelling conditions to improve well-being (Ross et al. 2012). When it comes to both importance and implementation, practical factors that include performance, service standards, continuous improvement, efficiency, and cost-effectiveness are highly scored in the topic of housing repairs (Tucker et al. 2014).

Improvements to dwelling conditions typically result in higher values for homes and communities, affecting subsidized and non-subsidized homes with higher rent, with the latter having more consequences including more financial and dwelling condition issues (Ellen et al. 2016). For Canada, improvements to housing issues will need to acknowledge this risk while still committing to repaired and adequate homes. The planning and regulation of health and safety at workplaces have been given more attention by the state and public than housing, resulting in housing regulations in comparison that have developed slowly in England (Burridge and Ormandy 2007). Social housing providers in England are challenged by ensuring that funds are used appropriately, resulting in maintenance and repair strategies that are lean and efficient (Tucker et al. 2014). This can also be argued in other developed nations which support and offer social housing. It is argued that the ideal approach to improving dwelling conditions is through incumbent upgrading, which allows existing residents to improve their dwellings, and minimizes residential instability and disruption (Perkins et al. 2009).

New national housing strategies or policies have occurred in other developed countries, such as Australia, but have been lacking in certain areas such as housing-related inequalities, tenures, location, and housing quality (Milligan and Pinnegar 2010). The labor government of the United Kingdom introduced the Housing Act of 2004, which identified 29 potential housing hazards, its intensity regarding housing conditions, and how it should be dealt with, up from the nine building-related requirements from 1990 (Burridge et al., 2007). Research by Walters (2009) presented the statistical results from the American Housing Survey, with no strong support that subsidized homes depreciate in housing quality compared to their non-subsidized counterparts, likely due to recent revitalization programs. Counteracting poor conditions with enacted revitalization strategies by several levels of government and organizations have included subsidized home improvement loans in the State of Utah (Perkins et al. 2009). National housing improvements in Wales for dwelling conditions have shown to improve social and health outcomes, but only after years instead of months or weeks of progress (Poortinga et al. 2017). These upgraded dwelling conditions in Wales have resulted in an improved status of health due to better thermal living conditions, and solutions associated with insulation, heating, and ventilation (Poortinga et al. 2017). These cases of regional and national housing strategies should be seen as examples that can be built on to improve housing plans for Toronto and Canada as a whole.

6.3 Data and Methodology

6.3.1 Selecting and Transforming Census Data from Statistics Canada

Data for this research was collected from Statistics Canada, downloaded at the census tract level in the City of Toronto and in the census years of 1996, 2001, 2006, 2011, and 2016. Arguably, this open data or accessible, transparent, and accountable data from the government should be used for the sake of empowering ordinary citizens and improving the understanding of government work in society (Baculi et al. 2017). Informed by the literature review, specific census profiles retrieved for the research included population and dwellings, age and sex, dwelling characteristics, income, labor, visible minorities, immigration, and mobility. Within these census profiles, individual census variables were selected and tested to be included as independent census variables. The census variables or predictors can be understood in two key groups regarding home and resident characteristics.

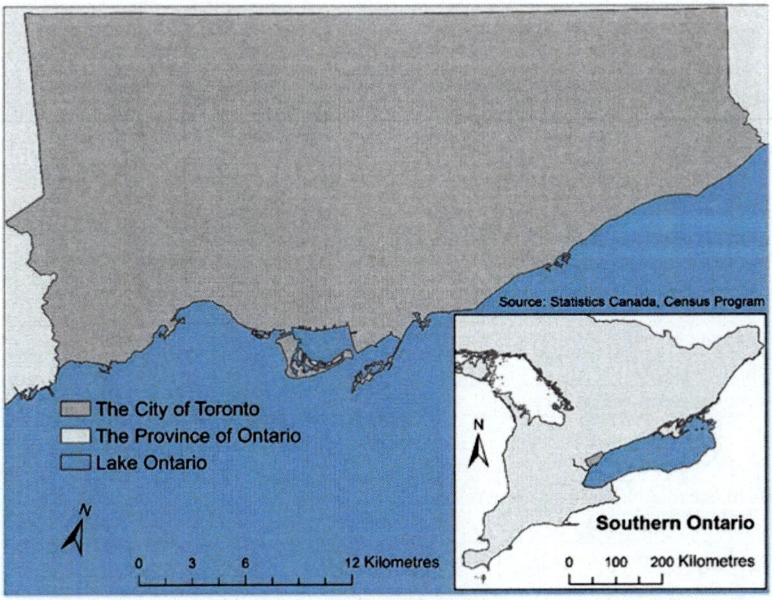

Fig. 6.1 Study area—the current borders of the City of Toronto

Thanks to insight from the literature review, combined with reports from organizations like United Way indicating home and resident characteristics, a number of these predictors could have a significant relationship with inadequate dwellings in Toronto. The data and potential predictors in question will eventually be visualized within the current borders of the City of Toronto, which can be found in Fig. 6.1. A total of 18 predictors for the independent variables were identified, extracted from the census, grouped appropriately for statistical analyses, and transformed to percentages. Criteria for the 18 predictors involved identifying home and resident characteristics that were significant in the literature and could be extracted and organized from the census data of Statistics Canada. Since a part of the research includes comparing statistical and spatial outputs over the census years of 1996, 2001, 2006, 2011, and 2016, the 18 predictors had to be available and comparable for those years. Table 6.1 presents the complete list of potential independent variables based on characteristics that were deemed important, were contributing, and had a relationship with dwelling conditions, primarily inadequate dwelling conditions.

These home and resident characteristics arguably have a position leaning toward vulnerable or disadvantaged groups and individuals, which is understandable given the focus on dwellings in need of major repairs. These independent census variables will be used in the analyses for the dependent variable of occupied private dwellings by dwelling condition with major repairs needed. This dependent variable is seen as an indicator of an inadequate dwelling. Together these independent and dependent variables will be used in the regression model for the census years of interest by census tract. In the bivariate analysis, the census variable of dwellings in need of

Table 6.1 Transformed census data—data derived from Statistics Canada and potential independent variables for statistical analyses

Home characteristics
Dwellings constructed in the past 20 or 25 years[a]
Dwellings constructed over 20 or 25 years ago[a]
Average number of rooms per dwelling
Apartment in a building that has five or more floors
Tenant households spending 30% or more of its income on shelter costs
Resident characteristics
Residents aged 0– 14[a]
Residents aged 65 and over[a]
Unemployment rate
Mobility—Moved in the past year
Mobility—Moved in the past 5 years
Household occupied by three or less residents[a]
Household occupied by four or more residents[a]
Visible minority population
Immigrant population
Married or living common law[a]
Not married and not living common law[a]
Language spoken most often at home—Nonofficial languages
Household income under $50,000[a]

[a]A derived variable aggregated from a number of variables

major repairs will be analyzed with the percentage of tenant households in subsidized housing as well as a derived variable of tenant households in non-subsidized housing. The dwelling condition regression residuals will also be used in a bivariate analysis with tenant household subsidy status.

6.3.2 *Methodology*

The completion of the cleaning, aggregation, and transformation of the census data progresses the study to the statistical analyses in the research. This is important since raw data may lead to misinterpretation of findings and possible distortions and miscommunication and facilitate user comprehension of the results (Wen and Zhou 2008). This normalization or adjustment of values of differing scales to a common scale allows analyses and comparisons to proceed. The 18 potential independent variables were processed in a hierarchal cluster analysis, and based on clustering, certain variables were used in a stepwise regression model.

A hierarchical cluster analysis consists of a repetitive partitioning of observations in a bottom-up or top-down approach to clustering, which results in a dendrogram with a clustering of data objects at similar levels compared to other data object clusters (Rokach and Maimon 2005). Visual insight is gained from the dendrogram and can be described as a tree graph that displays the clusters that have merged, with different levels and breaks indicating clustering (Schonlau 2004). A stepwise regression analysis consists of a model that automatically selects in a stepwise manner, involving partial correlations of the dependent variable against the independent variables that are close to ideal in the sense of exploiting the squared multiple correlation coefficient of the dependent variable against the chosen independent variables (Liu et al. 2007).

There are three model-selection approaches to stepwise regressions: forward selection, an empty model which adds variables that improve the model the most until ideal; backward elimination, a full model which removes the least improving variable until ideal; and bidirectional elimination, a forward selection with possible deletion of variables later (Wang et al. 2016). Using SPSS this study will use bidirectional elimination, to test for correlations between variables. Residuals were produced from the regression models for spatial analyses and compared to single census variable thematic maps. A bivariate analysis between dwelling conditions, its residuals, and tenant household subsidy status will complete the study. Figure 6.2 illustrates the methodology.

6.3.2.1 Hierarchal Cluster Analysis

Using the Statistical Package for the Social Sciences (SPSS), the spatial statistical research begins with the use of a cluster analysis, which is typically used to find the right number of clusters using a K-means algorithm (Vaz et al. 2016). In the case of this study, clusters of census variables are of interest and the number of produced clusters is of less interest. The cluster analysis will be used to find high homogeneity within the ideal cluster (of variables) while also having high heterogeneity from others (not ideal clusters of variables) (Spetic et al. 2008). This cluster analysis will be a hierarchical cluster analysis with squared Euclidean distances and supplemented by using a dendrogram to help identify the ideal cluster of variables (Cronley et al. 2018).

Like the research of Yoon (2015), the dendrogram will be instrumental in identifying clusters, similarities, and differences by cluster distances, and revealing which variables were closely related (along with the proximity matrix). The arrangement and generated clusters will not focus on the geographic distribution of the data, but rather the source of the variables and the groupings present in the dendrogram (Zimisuhara et al. 2015). The proximity matrix, which computes distances between objects, cases, and variables, will offer values that help explain why certain clustering had occurred and was present in the dendrogram (van der Kloot et al. 2005).

In this research, the 18 potential independent census variables of interest will be processed in SPSS for a hierarchical cluster analysis. Based on the resulting proximity matrix and dendrogram, clusters will be created, understood, and selected for the next statistical tests. The selected cluster and the variables present in the cluster

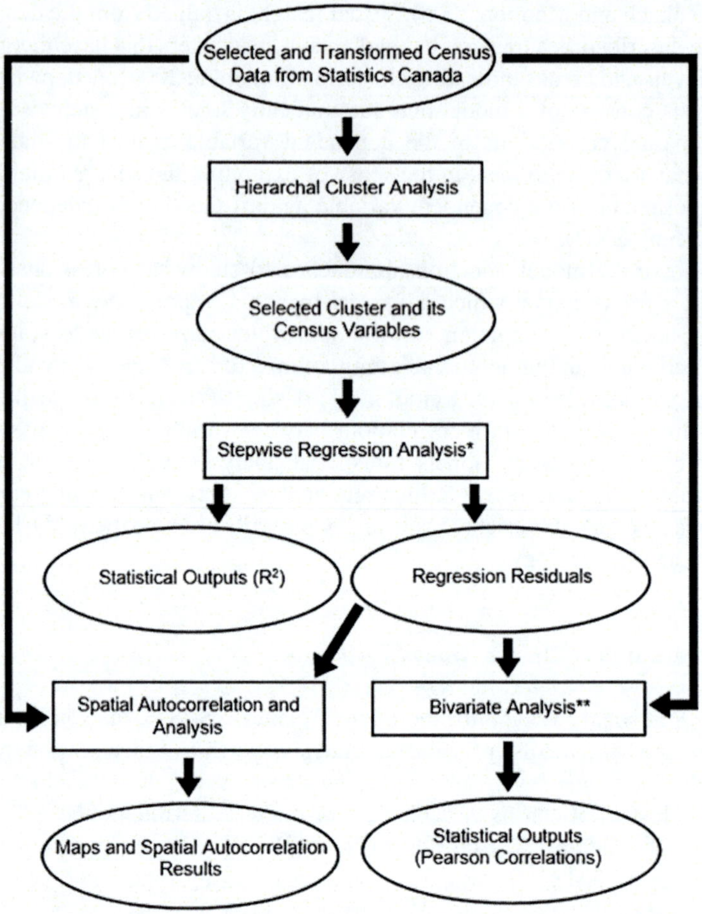

Fig. 6.2 Methodology flowchart *Repeated for the census years from the 2016 results, **Using only 2016 data

will be used in a stepwise regression model, which will be the main explanatory model for the research.

6.3.2.2 Stepwise Regression Analysis

The completion of the cluster analysis will follow a stepwise regression model to produce statistical results and highlight significances and residual values by census tracts for maps. Research by Wright and Kloos on the nature of housing environment influences used regression models to test hypotheses and show significance (2007). The independent variables for the regression model will be based on previous studies that affect housing, such as affordability and adequacy, and will be on household attributes and housing attributes (Lin et al. 2014).

The regression model will take the independent census variables as predictor variables for the dependent variable of interest (Shier et al. 2016). Research by Lee, Parrott, and Ahn also used dwelling conditions or housing adequacy as the dependent variable, where a binary variable coded as adequate or inadequate was made and will be partially similar to the single dependent variable of this research of "dwelling conditions in need of major repairs" (2014). Achieving a final regression model or explanatory model is possible with the use of a stepwise logistic regression with significant variables (Patterson et al. 2012). The use of a stepwise regression ensures the selection of variables with importance that improves the robustness of the procedure and increases the R^2 (Wang et al. 2016).

From the outcome of the hierarchical cluster analysis, numerous census variables from the selected cluster will be entered into a stepwise regression model using SPSS. The first model will use the 2016 data as a guinea pig to inform the models of the previous census years. As the recently conducted census, the outcome and significant relationships found would be relevant to the present day. The stepwise regression model will produce a list of census variables that have been entered, forming an explanatory model. When the list of predictors is established, these variables will be selected in the previous census years. This will result in the opportunity to compare the variance inflation factor (VIF), results of the Durbin-Watson test, the standard coefficients beta, the Pearson correlation values, and the R^2 of the census years of interest over time. Along with these statistical findings, the resulting residuals will be entered in GIS to create maps of these explanatory models over time and will be used in the bivariate analysis.

6.3.2.3 Bivariate Analysis

Informed by the completed regression or explanatory model, a bivariate analysis will be used between the 2016 census variable on dwellings in need of major repairs and its residuals and on tenant subsidized dwellings as well as a derived variable of tenant non-subsidized dwellings. In particular, the values of the Pearson correlations will inform the research of positive or negative correlations, and how significant the relationships were among the variables.

Research by Rosenbaum and Friedman also used bivariate analyses between housing conditions and social predictors, such as racial factors (2004). Measures of inequalities—or in the case of this research, a type of dwelling condition—will be calculated to evaluate differences among social groups, which in this case will be tenant dwellings which are subsidized or not, using SPSS (Roy et al. 2014). Work will be similar to the research by Kohlhuber, Mielck, and others which assessed how and whether relationships varied between housing conditions and other variables through a bivariate analysis (2006). Dwelling ownership and tenant status have been used in other studies with bivariate analyses and as dependent variables with independent variables regarding housing and household well-being (Dyb 2016).

The previous explanatory model from the stepwise regression will inform the outcome of the bivariate analysis, offering predictors with a significant relationship to inadequate dwellings. The literature review provides some insight on whether the chance of an inadequate dwelling worsens due to tenant subsidy status, with points for or against the likelihood of poorer conditions of one over the other. The recent status of dwelling condition in the City of Toronto based on the 2016 census will be fully understood with the completion of the bivariate analysis, alongside the findings of the 2016 census regression results.

6.3.2.4 Spatial Autocorrelation and Spatial Analysis

From the transformed census data and the residuals from the stepwise regression models of the recent census years, numerous maps were created to understand the spatial patterns in the city. Clusters present were validated using the spatial autocorrelation tool in ArcGIS to confirm spatial insights. Ensuring spatial autocorrelation is important since it answers the fundamental question to geography, of whether a spatial pattern of something is significant in some sense and worth understanding (Getis 2007).

The spatial autocorrelation tool will be used on both the single census variable maps on inadequate dwellings and their regression residual counterparts of their respective census years. Once spatial autocorrelation has been recognized in the data and corresponding maps, an investigation into where certain patterns and extreme values are present in the city can proceed. Spatial insights by census tracts would be compared between residual and non-residual maps, but perhaps more importantly among the census years to understand the changes or evolution of inadequate dwellings and its predictors in the city. Together the statistical and spatial analyses create an understanding of inadequate dwelling conditions and will be very informative for interventions such as a National Housing Strategy. Relationships and circumstances will be discovered that can educate decision-makers in targeting investments and programs, all for the sake of improving dwelling conditions, particularly in the rapidly changing urban environment of Toronto (Vaz and Arsanjani 2015).

6.4 Results

6.4.1 Hierarchal Cluster Results

The results of the hierarchal cluster analysis were produced by inputting 18 transformed census variables through SPSS from the 2016 census. Informed by the dendrogram of Fig. 6.3, it can be observed in the dendrogram that 17 of the 18 selected census variables have clustered together, while one single census variable is

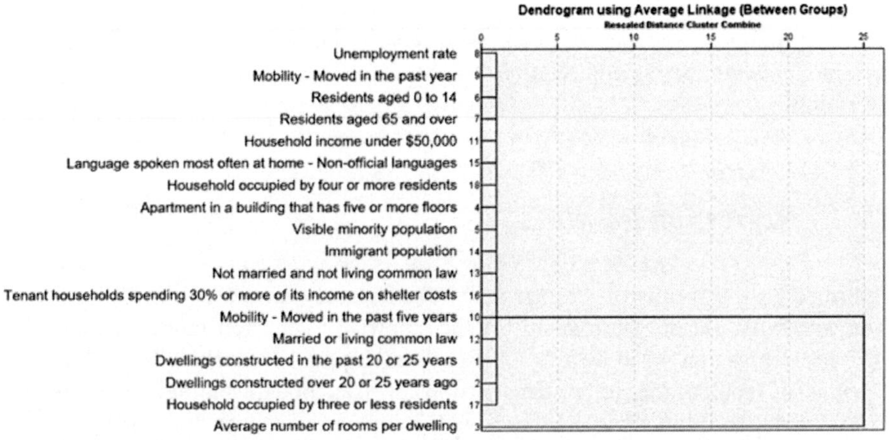

Fig. 6.3 Hierarchal cluster analysis—dendrogram results

clustered alone. The lone census variable of "average number of rooms per dwelling" had a significant distance of a linkage between its group and the other larger cluster.

This observable, considerable distance linking the two produced clusters notifies a similarity among the 17 census variables, but exceptional distance to the cluster with a lone census variable. Understanding why this occurred can be explained by the proximity matrix values produced in the hierarchal cluster analysis. Among the 17 variables in the initial cluster, the majority of the values produced were high values nearing 1.0. In the case of the variable of "average number of rooms per dwelling," the proximity matrix values produced in relation to the other census variables were below 0.2. These extremely low values in comparison to the other proximity matrix values allow additional understanding of the dendrogram created in the analysis.

The 17 census variables from the cluster of interest include unemployment rate, mobility—moved in the past year, residents aged 0 to 14, residents aged 65 and over, household income under $50,000, language spoken most often at home—non-official languages, household occupied by four or more residents, apartment in a building that has five or more floors, visible minority population, immigrant population, not married and not living common law, tenant households spending 30% or more of its income on shelter costs, mobility—moved in the past 5 years, married or living common law, dwellings constructed in the past 20 or 25 years, dwellings constructed over 20 or 25 years ago, and household occupied by three or less residents.

This single cluster containing 17 census variables, based on findings from the proximity matrix and dendrogram, was selected for entry into a stepwise regression model. The second cluster with only the "average number of rooms per dwelling" variable was not analyzed further in the regression analysis. The completion of the hierarchal cluster analysis progresses the research by identifying a group of significant census variables with considerable similarities. This also narrows down the list of potential independent census variables, in order to improve the explanatory model when identifying relationships and circumstances regarding inadequate dwellings.

6.4.2 Regression Results

Informed by the results of the hierarchal cluster analysis, the 17 census variables from the 2016 census were taken for the stepwise regression model. This initial stepwise regression acted as a "guinea pig" test using 17 variables from the 2016 census. The result of the stepwise regression consisted of a lower number of selected census variables from 17 variables, as well as numerous values regarding the standard coefficients beta, the VIF, the Pearson correlation values, results of the Durbin-Watson test, and the R^2. The selected census variables from this regression were then selected in the other census years of interest of 2011, 2006, 2001, and 1996. These additional regression models would mirror the initial regression analysis of the 2016 census, producing statistical findings and residuals that would be compared for this study.

From the initial 17 census variables, eight variables or predictors were selected and entered in the model: dwellings constructed over 20 or 25 years ago, apartment in a building that has five or more floors, residents aged 0–14, residents aged 65 and over, mobility—moved in the past 5 years, household income under $50,000, language spoken most often at home—nonofficial languages, and tenant households spending 30% or more of its income on shelter costs. These eight predictors were then selected in the previous census years of interest, forming an explanatory model, based on the past 20 years by census years. The results of these models can be understood by comparing key statistical outputs over the census years. As stated earlier, these key statistical outputs include the standard coefficients beta, the VIF, the Pearson correlation values, results of the Durbin-Watson test, and the R^2. From the values produced, patterns and insight on the relationships between the dependent variable of dwellings in need of major repairs and inadequate dwelling predictors can be analyzed.

Table 6.2 displays the standard coefficient beta values produced from the stepwise regression analysis of the 2016 census and repeated with the same independent census variables of the previous census years. When examining the values or standard deviations produced, the strength of each independent variable to the dependent variable of dwellings in need of major repair was explored. The exploration of these coefficients allows for the comparison of relative importance.

The table below reveals that of the eight selected independent census variables, two of these predictors have a significantly higher and therefore stronger effect on the dependent variable. The predictor "dwellings constructed over 20 or 25 years ago" had the second highest standard deviations over the course of the past 20 years by census year. The highest standard deviations were found among "household

Table 6.2 Standard coefficients beta—stepwise regression results

Standard coefficients beta	Census year				
Inadequate dwelling predictors	2016	2011	2006	2001	1996
Dwellings constructed over 20 or 25 years ago	0.360	0.394	0.431	0.519	0.398
Apartment in a building that has five or more floors	−0.093	−0.069	−0.100	−0.084	0.015
Residents aged 0–14	0.157	0.097	−0.010	0.026	−0.051
Residents aged 65 and over	−0.127	−0.078	−0.202	−0.103	−0.204
Mobility—Moved in the past 5 years	−0.104	−0.059	−0.099	0.160	−0.039
Household income under $50,000	0.764	0.645	0.584	0.376	0.520
Language spoken most often at home nonofficial languages	−0.215	−0.162	−0.115	−0.071	−0.145
Tenant households spending 30% or more of its income on shelter costs	−0.063	−0.100	0.027	−0.137	−0.007

Table 6.3 Variance inflation factor—stepwise regression results

Variance inflation factor (VIF)	Census year				
Inadequate dwelling predictors	2016	2011	2006	2001	1996
Dwellings constructed over 20 or 25 years ago	1.981	1.914	1.982	1.850	2.157
Apartment in a building that has five or more floors	2.781	2.375	2.764	2.157	2.483
Residents aged 0–14	1.528	1.514	1.601	1.639	1.867
Residents aged 65 and over	1.672	1.507	1.760	1.755	2.066
Mobility—Moved in the past 5 years	2.094	1.940	1.670	1.579	2.023
Household income under $50,000	3.040	3.159	5.709	3.479	3.973
Language spoken most often at home nonofficial languages	1.427	1.429	1.642	1.590	1.711
Tenant households spending 30% or more of its income on shelter costs	1.215	1.238	5.918	1.296	1.387

income under $50,000," which indicates that this predictor has the strongest effect on inadequate dwellings. It should be noted that this predictor regarding income mostly increases in the past 20 years. This supports findings in the literature that income and the dwelling age proxy variable have the strongest relationship with inadequate dwellings, among the eight significant predictors.

Table 6.3 displays the variance inflation factor (VIF) values produced from the stepwise regression analysis of the 2016 census and repeated with the same independent census variables of the previous census years. In the regression analyses conducted, the produced VIF values allow the study to detect multicollinearity. Since the research had a focus on identifying significant and independent predictors, it should be ensured that relatively low correlation or correlation of little concern is found among the selected census variables.

The table below reveals that over the years, low to moderate correlations were found between the predictors and the dependent variable of inadequate dwellings. In the 2006 census, the highest VIF values that were above 5.0 were found with "household income under $50,000" and "tenant households spending 30% or more of its income on shelter costs," which are the highest VIF values present. Although

Table 6.4 Pearson correlation values—stepwise regression results

Pearson correlation values	Census year				
Inadequate dwelling predictors	2016	2011	2006	2001	1996
Dwellings constructed over 20 or 25 years ago	0.612	0.609	0.652	0.601	0.545
Apartment in a building that has five or more floors	0.281	0.266	0.265	0.222	0.234
Residents aged 0–14	−0.062	−0.104	−0.157	−0.105	−0.190
Residents aged 65 and over	0.066	0.120	0.098	0.072	0.093
Mobility—Moved in the past 5 years	0.019	0.061	−0.013	0.150	0.192
Household income under $50,000	0.648	0.600	0.630	0.506	0.572
Language spoken most often at home nonofficial languages	−0.058	0.030	0.093	0.122	0.148
Tenant households spending 30% or more of its income on shelter costs	−0.005	0.116	0.557	0.057	0.177

these are the highest correlations present in the research, these values are accepted with little concern since the VIF is below 10, which minimizes a concern for multicollinearity. Since there is low to moderate correlation among the predictors, the issue of multicollinearity throughout the numerous regression models which form the explanatory model is minimal. These findings support all eight predictors as independent variables to inadequate dwellings.

Table 6.4 displays the Pearson correlation values produced from the stepwise regression analysis of the 2016 census and repeated with the same independent census variables of the previous census years. These correlation coefficients were used to measure the strength of the relationship between the dependent variable of inadequate dwellings and the eight selected independent predictors. These values allow the identification of positive or negative correlations along with the strength of the relationships.

The table below reveals that of the eight predictors, four census variables have values that stand out over the past 20 years. The predictor "residents aged 0 to 14" had a consistently negative correlation with the dependent variable. This suggests that the instance of children in the household slightly lowers the chance of an inadequate dwelling in Toronto; however, the values also suggest low correlation. "Apartment in a building that has five or more floors" had the third highest values with a positive correlation, with consistent values above 0.2. The two predictors with the highest values were "dwellings constructed over 20 or 25 years ago" and "household income under $50,000." This once again supports the notion that income and the dwelling age proxy variable have a significant and positive relationship with inadequate dwellings, along with the "apartment in a building that has five or more floors" predictor.

This explanatory model had also produced Durbin-Watson test values which can be found in Table 6.5. The values found in the Durbin-Watson test are important since it tests for autocorrelation in the regression analyses. It is generally accepted that normal test statistic values ranging from 1.5 to 2.5 are relatively normal and

Table 6.5 R^2, Durbin-Watson and percentages—stepwise regression results and others

Census year	Toronto dwellings in need of major repairs	Average % of dwellings in need of major repairs	R^2 (R-squared)	Durbin-Watson test result
2016	5%	6.4%	0.683	1.807
2011	6%		0.569	1.769
2006	6%		0.610	1.853
2001	7%		0.481	1.872
1996	8%		0.479	1.884

should not be a cause for concern. In the case of this study, all the values range from 1.7 to 1.9, which means positive autocorrelation is present, with values that are acceptable and of no concern.

The R^2 in Table 6.5 informs the research of how well the model explained the variability of the response data around the mean. Over the past 20 years by census year, it can be observed that R^2 had increased toward the recent 2016 census. An exception to this is the R^2 of the 2011 census. This progression of an R^2 over the census years suggests that the relationship between the eight predictors and the dependent variable regarding inadequate dwellings had improved, resulting in a stronger relationship. An interesting side to this research was the findings that the percentage of inadequate dwellings had been decreasing over the census years. Together with the R^2, these findings inform us that while Toronto lowered its percentage in inadequate dwellings, the relationship to key census variables improved in the models. This suggests that in Toronto, a small group of dwellings have consistently experienced poor housing adequacy, in a relationship that had strengthen based on the eight predictors over time.

6.4.3 *Bivariate Results*

Thanks to the findings of the explanatory model based on the regression models, the bivariate analyses were informed of census variables which suggest significant relationships with inadequate dwellings. Along with an interest in inadequate dwellings, the research also investigated tenant housing subsidy status from the 2016 census. For this study, the regression residuals and the single census variable regarding inadequate dwellings were analyzed against the values regarding tenant housing subsidy status. The statistical outputs of interest were the correlation coefficients, which informs the study of the strength of the relationship between entered values.

Table 6.6 displays the correlation coefficients between the residuals and tenant subsidy status. These values show an insignificant to no relationship. Table 6.7 shows correlation coefficients with moderate relationships between inadequate dwellings and tenant housing subsidy status. These correlations are also important since they are significant at the 0.01 level.

Table 6.6 Pearson correlation values, residuals against tenant subsidy status—bivariate results

Pearson correlation	Subsidized tenant homes	Non-subsidized tenant homes
Regression model residuals (inadequate dwellings) 2016	−0.067[a]	0.067[a]

[a]Correlation is significant at the 0.05 level (two-tailed). No or negligible relationship

Table 6.7 Pearson correlation values, inadequate dwellings against tenant subsidy status—bivariate results

Pearson correlation	Subsidized tenant homes	Non-subsidized tenant homes
Dwelling in need of major repairs (inadequate dwellings) 2016	0.345[a]	−0.345[a]

[a]Correlation is significant at the 0.01 level (two-tailed). Moderate positive/negative relationship

These findings between inadequate dwellings and tenant housing subsidy status are arguably supported by the explanatory model results. In particular, subsidized homes have a significant moderate positive relationship, suggesting that an increased instance of subsidized homes is likely to result in a census tract with a higher instance of inadequate dwellings. Non-subsidized homes have the opposite result, with a significant moderate negative relationship with inadequate dwellings. The explanatory model with the eight selected significant predictors supplements the bivariate findings by providing statistical context. Given the nature of subsidized homes from the literature review, it is suggested that subsidized homes have a stronger relationship with the predictors identified in the explanatory model.

6.4.4 *Spatial Autocorrelation and Spatial Findings*

Exploring the geographic side of this spatial statistical research was achieved by transforming and creating geospatial data from the census data and the residuals produced from the numerous regression models. Along with making maps to visualize the data, the data underwent spatial autocorrelation to validate clusters present in the data and the eventual spatial insights found in the maps. Investigating the regression residuals was important since validating the outputs through spatial autocorrelation would support the statistical findings in the geographic context. This can also be said for the maps of the dependent variable, and together the residual results can support the findings from a geographic perspective.

Table 6.8 shows clustering to be present in all the census years, with the exception of the earliest census year in the study of 1996. This result of a random pattern for the 1996 census year is due to the z-score, p-value, and Moran's index values produced. This can also be supported by the R^2 value produced for 1996, which showed an R^2 of 0.479, the lowest R^2 produced of the census years. The other census years show values which indicate clustering. Arguably, these results are comforting

Table 6.8 Spatial autocorrelation results: clustering—regression residuals

Census year	z-score	p-value	Moran's index	Type of pattern
2016	1.906245	0.056618	0.026431	Clustered[a]
2011	2.670710	0.007569	0.043835	Clustered
2006	6.742170	0.000001	0.115303	Clustered
2001	5.144537	0.000001	0.088080	Clustered
1996	0.945027	0.344645	0.013022	Random

[a]There is a less than 10% likelihood that this clustered pattern could be the result of random chance

Table 6.9 Spatial autocorrelation results: clustering—census, inadequate dwellings

Census year	z-score	p-value	Moran's index	Type of pattern
2016	12.784640	0.000001	0.186845	Clustered
2011	10.325749	0.000001	0.174776	Clustered
2006	14.874786	0.000001	0.256852	Clustered
2001	13.459759	0.000001	0.233762	Clustered
1996	8.356297	0.000001	0.132358	Clustered

given that the regression residual maps are dubious in displaying clusters. These regression residual maps were beneficial in not only identifying patterns in the inadequate dwelling maps but also offering a level of confidence between the eight selected predictors and the dependent variable. Residuals which do not display an extreme over- or underprediction of the dependent variable would suggest a stronger level of confidence in the clusters or spatial patterns in the inadequate dwelling maps.

Table 6.9 shows the spatial autocorrelation results of the inadequate dwelling data and their maps. All the census years of the study show clustered patterns, indicating a high level of confidence in the clustering or spatial patterns observed in the maps. Although all the census years show clustering, when it comes to the 1996 census, the lowest z-score and Moran's index values are present. This suggests that along with the regression residual results, the R^2, and percentage of inadequate dwellings over the years since the 1996 census year, there has been a concentration and strengthening of the relationship between inadequate dwellings and its predictors in the City of Toronto over time.

When comparing the 1996 maps in Fig. 6.4, there is an acknowledgment that the spatial autocorrelation has indicated that a random pattern is present in the map of regression residuals. What can be observed in the residual map is the presence of residuals by census tract throughout the city which do not have extreme over- or underprediction by the dependent variable of inadequate dwellings. This suggests that the majority of spatial patterns present in the dependent census variable map have a decent level of confidence between them and the eight predictors identified in the research.

Exploring the spatial patterns in the 1996 map of inadequate dwellings, there are darker census tracts and therefore a higher presence of poor housing conditions moving toward the downtown core of the city. This cluster of inadequate dwellings

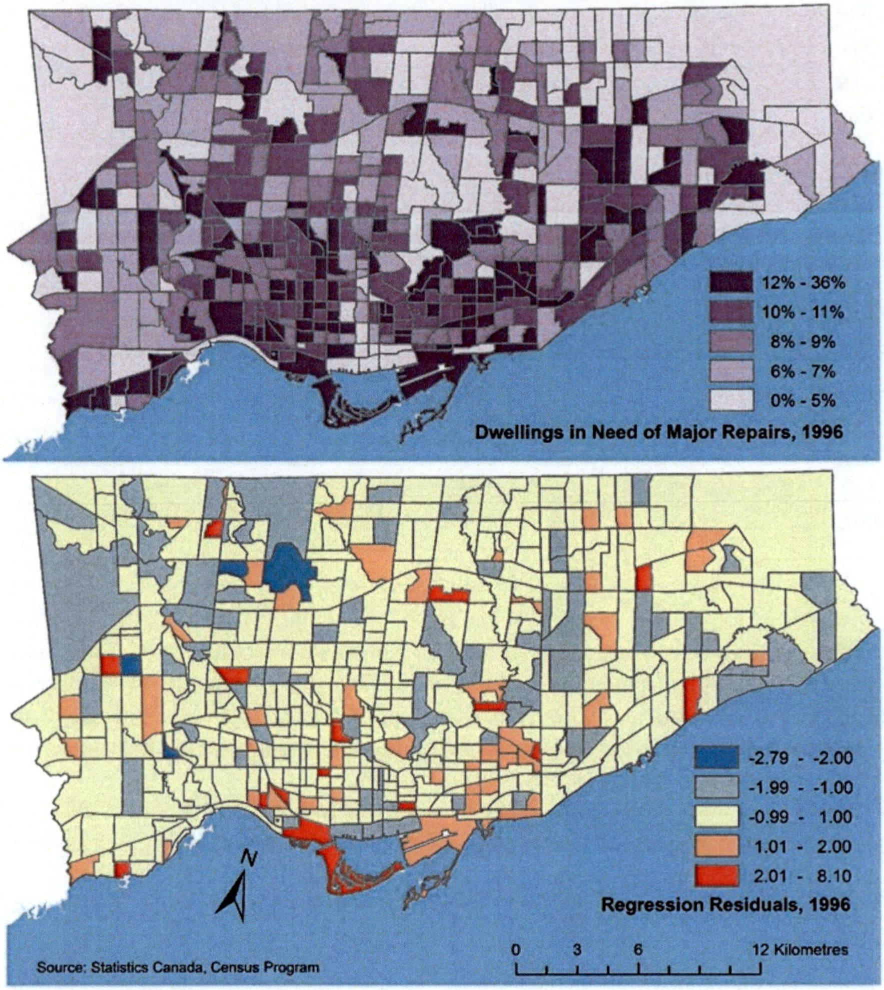

Fig. 6.4 1996—inadequate dwellings and its regression residuals, by census tract in the City of Toronto

can be seen in two waves: one in the southwest corner touching the downtown core and the other mirroring in the southeast. Lower instances of inadequate dwellings are found in the center of the city, the outer edges of the borders of Toronto, and the waterfront of downtown Toronto. Since household income under $50,000 was considered one of the most significant predictors in the research, it is likely that the alternative or higher-income households are playing a role in these areas. The nature of the eight predictors which are concerned with instances of high rent, lower incomes, older dwellings, the use of nonofficial languages, recent moving, the children, and elderly are likely found in the areas of high clustering. Based on the literature, these predictors suggest that vulnerable and marginalized communities in need of government assistance are likely to be housed in dwellings in need of major repairs. As the research explores the results of the next census years, the spatial patterns and strength of the relationship between the dependent variable and predictors are likely to become stronger and concentrated.

The 2001 census year carries the research over to the twenty-first century. In the 2001 census, the percentage of inadequate dwellings in the city dropped from 8% to 7%, while the R^2 had a slight increase from 0.479 to 0.481. The spatial autocorrelation tests also reveal that 2001 was the first census year in the residual maps to have a clustered pattern. These results and values suggest a start of the improved concentration of inadequate dwellings in the city, along with a slight strengthening of the relationship with the eight predictors. Similar to the 1996 regression residual map, the majority of census tracts by residuals appear to not have extreme over- or underprediction in the city. This finding suggests that spatial patterns found in the inadequate dwelling map have a confident relationship with the predictors.

Figure 6.5 illustrates the geographic findings of the 2001 census year. As identified in the 1996 maps, the two waves of inadequate dwellings are still present 5 years later in the city. The instance of dwellings in need of major repairs had slightly shifted north compared to the previous census year. Areas of the city with few instances of inadequate dwellings are relatively the same in the 2001 census year, with less clustering in the center of the city. However, it should be noted that this is represented by a slightly small change in percentage. Since one of the eight predictors of the research is "mobility—moved in the past five years," these slight shifts in percentage may be the result of slight changes among the census years due to moving. Comparing the 1996 and 2001 figures, it can be seen that the highest percentage of inadequate dwellings decreases from 36% to 29%. These census tracts are less dense and have a high quantity of park space. In the case of the 1996 census, the Toronto Islands had the highest instance of inadequate dwellings. This progresses the research by suggesting an improved focus on the highly urbanized census tracts. This also ensures more focus on highly residential areas with vulnerable and marginalized communities.

Moving on to the 2006 census year, the percentage of inadequate dwellings in the city had dropped again by 1% from 7% to 6%. The R^2 also increased from 0.481 to 0.610 and clear clustering is found in both the inadequate dwellings and residual maps based on spatial autocorrelation results. These results continue the trend of a concentration and yet strengthening of inadequate dwellings and their relationship with the eight predictors. Figure 6.6 displays the regression residual map with a majority of census tracts not exhibiting extreme over- or underpredictions. This suggests a good level of confidence between the predictors and the spatial patterns of the inadequate dwelling map.

The inadequate dwelling map in Fig. 6.6 compared to its 2001 predecessor shows a decrease in dwellings in need of major repairs. Like the previous maps, these areas of low percentage are found in the outer edges of the city, the waterfront, and the center of the city. The waves of inadequate dwellings are found on the southeast and southwest corners of the downtown core, stretching into the northeast and northwest corners of the city. As the third census year map in the study, the spatial patterns

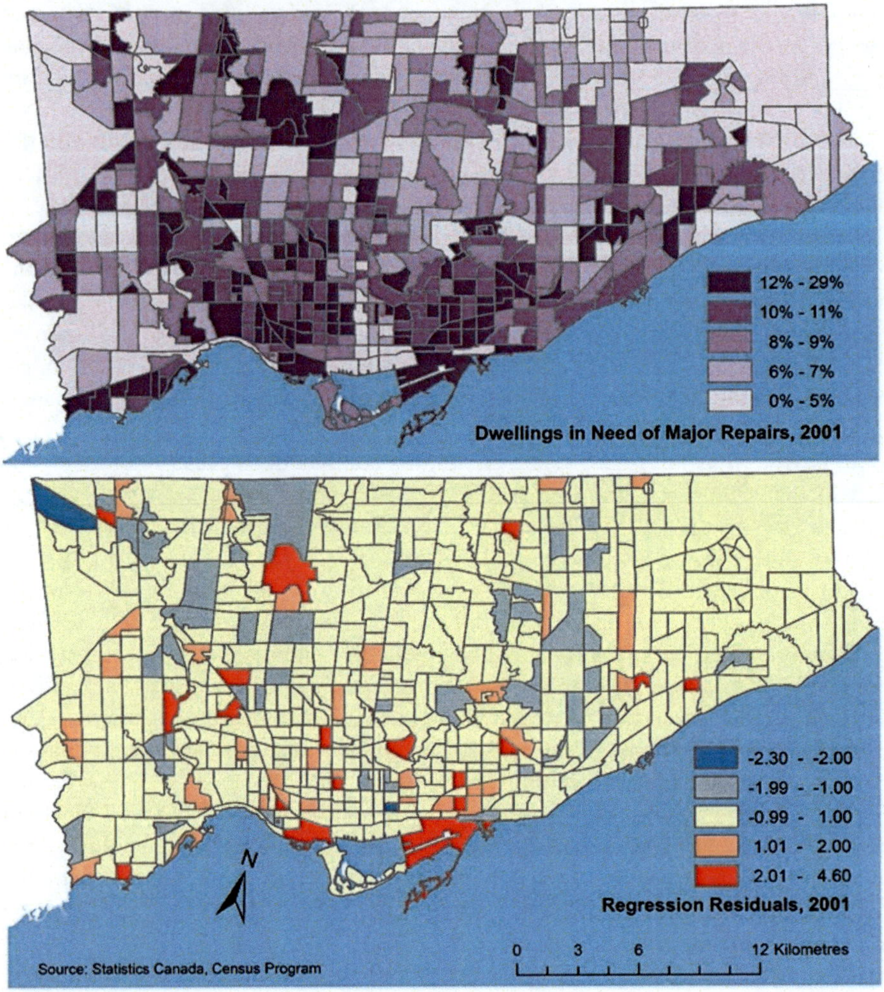

Fig. 6.5 2001—inadequate dwellings and its regression residuals, by census tract in the City of Toronto

present contribute to a trend of locations. By census tract and their nearby neighbors, these locations indicate slight changes in the percentage of inadequate dwellings as well as concentrations over time. In the areas of the city with high percentages of dwellings in need of major repairs, these areas appear to be consistent. This suggests that along with the residual maps, the relationship with the eight predictors were a factor in their results. These census tracts are also found much closer to the highly dense and urban downtown core. This suggests that higher populations are effected and are likely found in apartment buildings with five or more floors, one of the significant predictors.

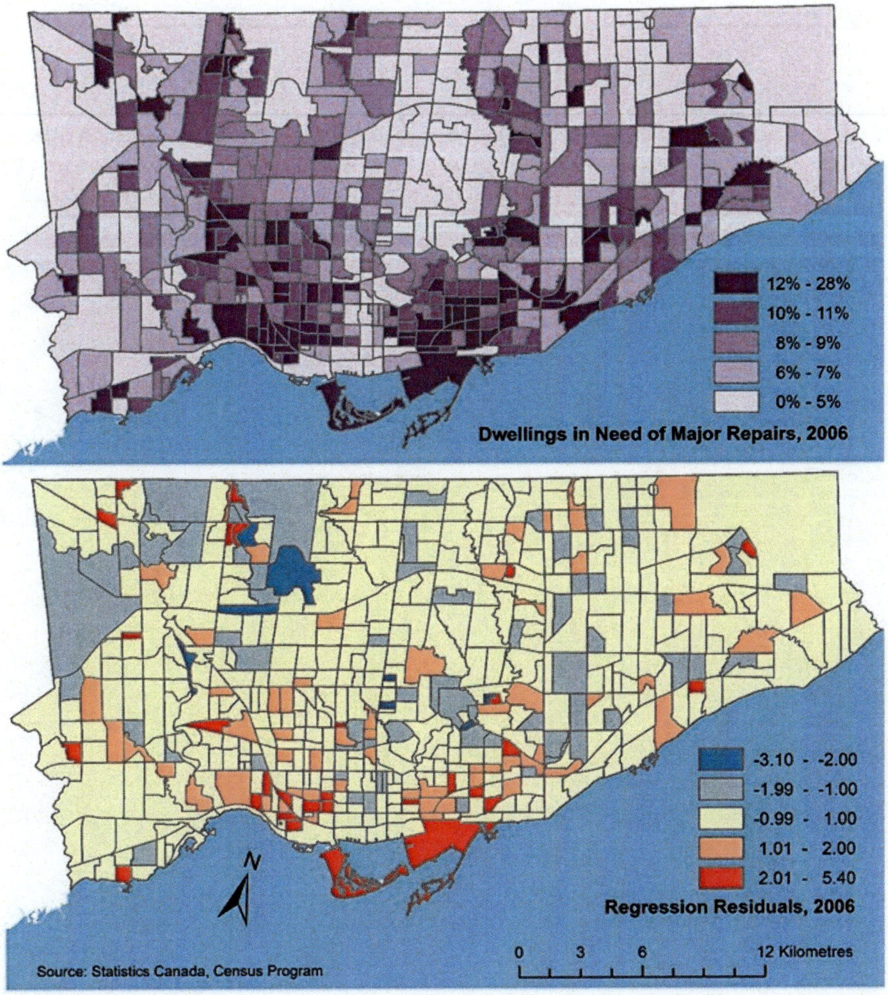

Fig. 6.6 2006—inadequate dwellings and its regression residuals, by census tract in the City of Toronto

The 2011 census year differed by having a stabilization of the percentage of inadequate dwellings in the city at 6%, the same percentage as the 2006 census year. There was also a lowering of the R^2 from 0.610 to 0.569. Although this indicates a slight weakening in relation to the eight predictors of the study, the relationships are still significant. Clustering was still present based on the spatial autocorrelation results of the inadequate dwelling and residual maps in Fig. 6.7. The 2011 results continue the trend of the regression residual maps, with the majority of census tracts not exhibiting extreme over- or underpredictions.

In terms of the 2011 inadequate dwelling map of Fig. 6.7, the spatial pattern or clustering of dwellings in need of major repairs from the previous census years has

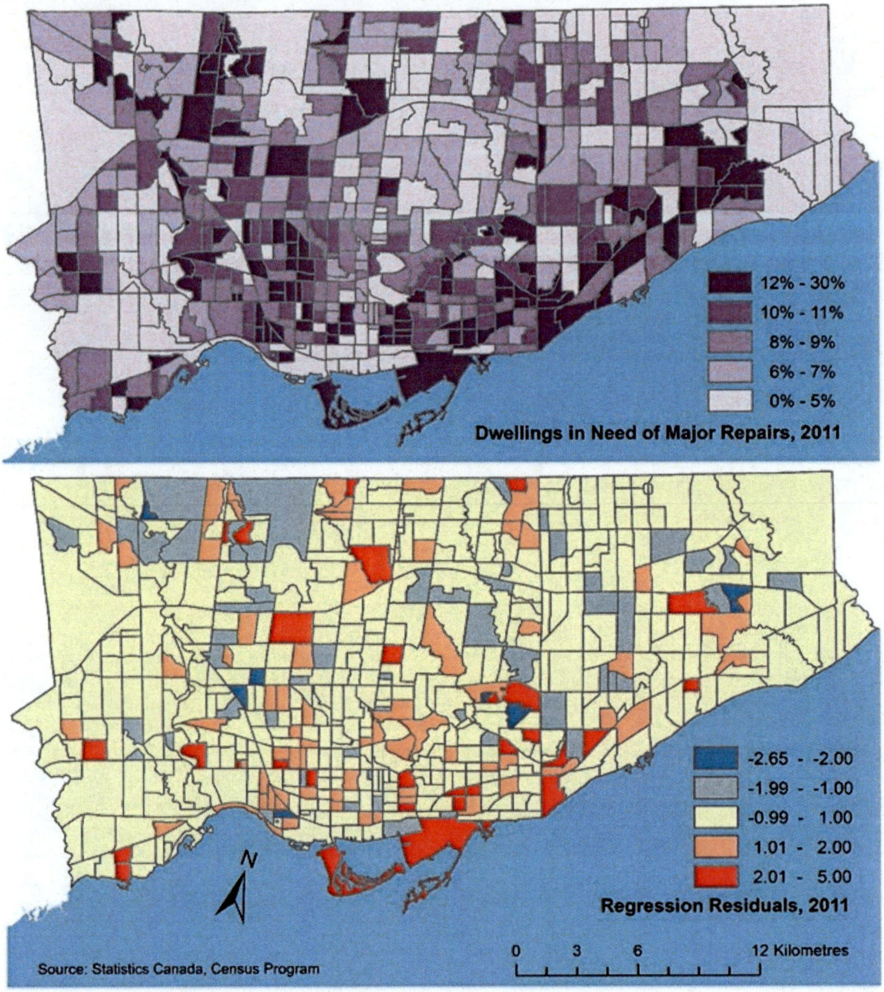

Fig. 6.7 2011—inadequate dwellings and its regression residuals, by census tract in the City of Toronto

continued. The location of higher and lower instances of inadequate dwellings in the city appears to be similar to the other census years. Even with slight color changes and therefore percentage changes by census tract from one census year to the next, the concentration of inadequate dwellings is still found on the southeast and southwest corners of the downtown core. This fourth census year contributes to the overall trend of where inadequate dwellings are present in the city.

The 2016 and final census year in this research experienced a drop in the percentage of inadequate dwellings in the city from 6% to 5%. The R^2 also jumped from 0.569 to 0.683, which was the highest R^2 recorded in the study. The recent census of

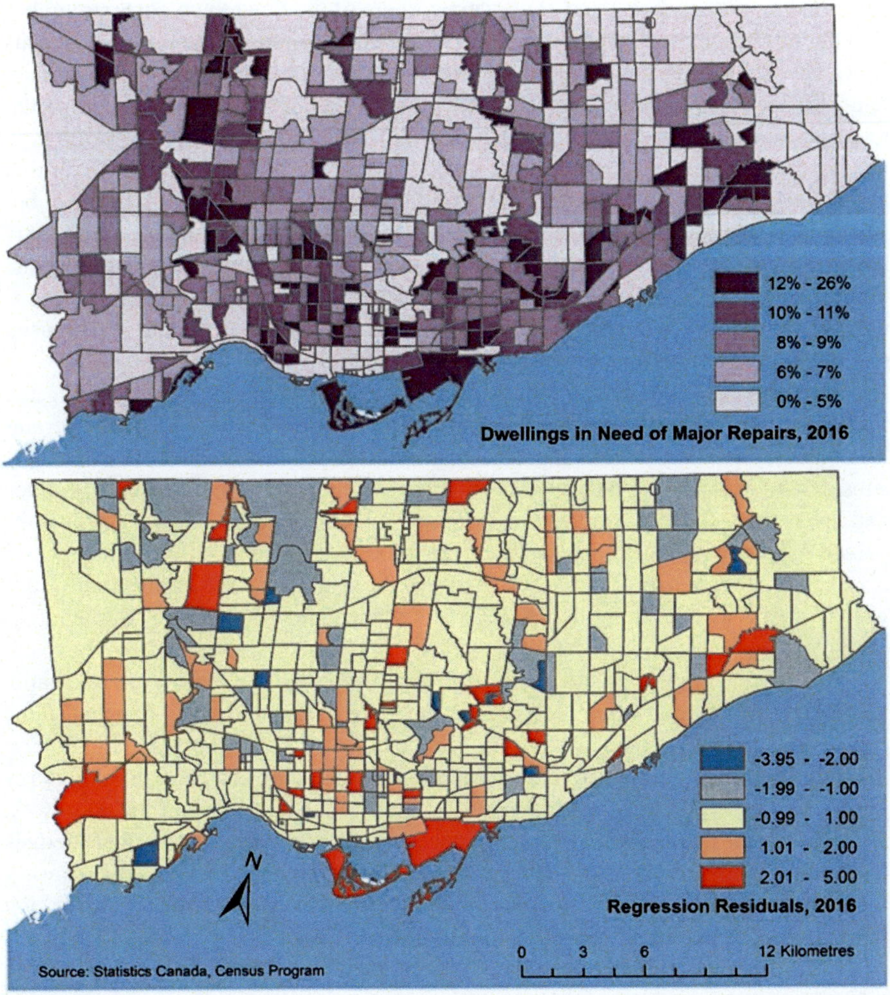

Fig. 6.8 2016—inadequate dwellings and its regression residuals, by census tract in the City of Toronto

Statistics Canada, along with the explanatory model of this study, informs the research that the relationship between inadequate dwellings and the selected eight predictors had recently strengthened. According to the spatial autocorrelation results, both the inadequate dwellings and residual maps had clustering. Like the previous census years, the majority of regression residuals by census tracts did not exhibit extreme over- or underpredictions (Fig. 6.8). Looking at all five census years in the research, on average 6.4% of the dwellings in Toronto were dwellings in need of major repairs, and as the years progressed, the percentage dropped over time.

Figure 6.8 finishes off the inadequate dwellings and regression residual maps with the 2016 census data. Once again the high percentage of inadequate dwellings are found on the southeast and southwest corners of the downtown core, stretching into the northeast and northwest corners of the city. Similar to the previous census years, the waterfront, the edges of the city, and the center of the city have consistently displayed lower percentages of inadequate dwellings. Together the completed five figures on the census years show a trend of clustering in regard to dwellings in need of major repairs. These statistical and spatial results inform the research that inadequate dwellings have experienced small decreases in the City of Toronto, but based on the explanatory model from the regression analyses, the eight predictors have strengthened its relationship with the dependent variable. The nature of the eight predictors, supported by the literature, indicates that communities or census tracts with high levels of marginalized and vulnerable communities are at increased risk of residing in dwellings in need of major repairs.

Understanding the bivariate analysis results can be supplemented by using the maps in Fig. 6.9. In the bivariate analysis results, focusing on the Pearson correlation values, the analysis informed the research that against inadequate dwellings, subsidized tenant homes had a value of 0.345, while non-subsidized tenant homes had a value of −0.345. Using Fig. 6.9, these Pearson correlation values are accompanied with a geographic context. Since the majority of residuals by census tract do not exhibit extreme over- or underpredictions, the spatial patterns of discussion suggest a good level of confidence between inadequate dwellings and the eight predictors.

When observing the spatial patterns in the dwellings in need of major repairs, suggested assumptions when compared to the maps of tenant subsidy housing status can be made. The Pearson correlation coefficient of 0.345 indicates a significant, moderate positive relationship between subsidized housing and the spatial patterns observed. These patterns, characterized by high values clustered in the southwest and southeast corners of the downtown core and the northwest and northeast edges of the city, are evident in both the inadequate dwelling and subsidized housing maps. Adding to the strength of this relationship, census tracts of the city which have no subsidized tenant homes appear to have low instances of inadequate dwellings. In the case of the non-subsidized housing map, not only is the opposite spatial pattern present compared to its counterpart, but its clusters are arguably inverted patterns of the inadequate dwelling map. These clusters help explain why certain Pearson correlation values were produced from the bivariate analysis. These findings suggest a positive correlation between dwellings in need of major repairs and subsidized tenant homes, and a negative correlation with non-subsidized tenant homes. It can be further argued that the eight predictors are more likely to be present among subsidized tenant homes due to the statistical and spatial findings.

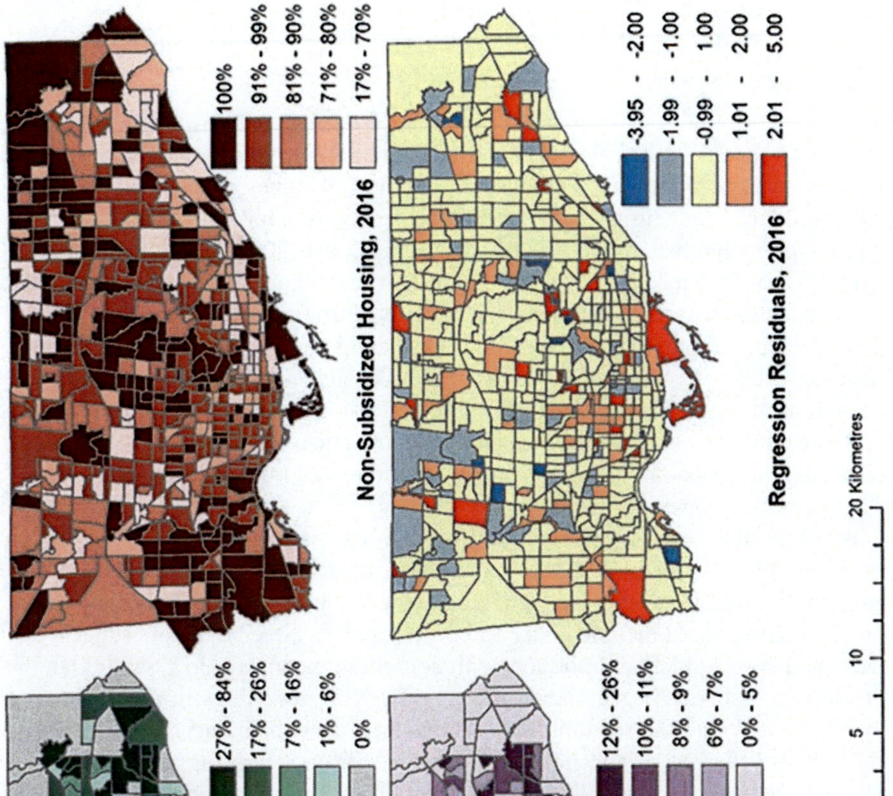

Fig. 6.9 2016—tenant subsidy household status, inadequate dwellings and its regression residuals, by census tract in the City of Toronto

6.5 Discussion

6.5.1 Regression Model Revelations

Attaining a spatial statistical understanding in this research was achieved by capturing the trend and significant findings. One of the key values of interest was the R^2 values produced per census year. The R^2 had a progression for the most part, increasing in value: 1996 had an R^2 of 0.479, 2001 had 0.481, 2006 had 0.610, and 2011 had 0.569, and the recent census of 2016 had an R^2 of 0.683. This suggests that the eight predictors over the years increasingly had a stronger relationship with inadequate dwellings in the City of Toronto. These eight census variables or predictors included dwellings constructed over 20 or 25 years ago, apartment in a building that has five or more floors, residents aged 0 to 14, residents aged 65 and over, mobility—moved in the past 5 years, household income under $50,000, language spoken most often at home—nonofficial languages, and tenant households spending 30% or more of its income on shelter costs.

Based on the Pearson correlations from the regression analyses, the following predictors over the years had the highest and most significant values: dwellings constructed over 20 or 25 years ago, apartment in a building that has five or more floors, and household income under $50,000. Together these predictors suggest that older, high-rise residential buildings with household incomes below $50,000 in the city have a statistically significant chance of having dwellings in need of major repairs. Joined with the remaining five predictors, the explanatory model comprising stepwise regression and iterations over the census years suggests that demographics that were marginalized and vulnerable in terms of home and resident characteristics would likely experience inadequate dwellings. These characteristics can be found in the literature and are highly correlated with homes in need of major repairs. Increasingly new immigrants are richer, highly educated, and bringing new businesses and have prosperous immigrant neighborhoods and the ability to choose homes that have adequate dwelling conditions (Carter 2005). This could explain why the census variables of immigrants and visible minorities were not selected.

6.5.2 Bivariate Relationships and Comparisons

The bivariate analysis produced a Pearson correlation value of 0.345 for the relationship between dwellings in need of major repairs and subsidized tenant homes, and a value of −0.345 against non-subsidized tenant homes. These correlations are significant at the 0.01 level and are also moderate relationships. These significant results along with the explanatory model findings appear to correlate subsidized tenant dwellings closely to the effects of inadequate dwellings. This notion is supported by the literature which states that marginalized and vulnerable populations are found in subsidized homes, and their characteristics are also found among those living in inadequate dwelling conditions. The literature also states that private non-subsidized tenant homes may also suffer from inadequate dwellings. Based on the statistical outputs, the results suggest different and private tenant management in Toronto may be a factor in ideal housing adequacy when compared to their subsidized counterparts.

It should be mentioned that the non-subsidized tenant dwelling variable was calculated based on the existing subsidized tenant dwelling variable. This explains why the results were the opposite of each other in the bivariate analysis, but still at a moderate level suggest that non-subsidized tenant dwellings have a better outcome in terms of dwelling condition. These results for the City of Toronto propose that certain home and resident characteristics are significant in instances of inadequate dwelling conditions. Adding the data regarding tenant household subsidy status brings more focus to the research indicating a stronger relationship with subsidized over non-subsidized tenant homes.

6.5.3 Spatial Patterns and Insights

In the previous numerous figures from the explanatory model, the inadequate dwelling maps display a clustering of high values by census tract at the corners of the downtown core spreading northeast and northwest of the city. This high value clustering is consistent and the low values at the center of the city, the waterfront, and the outer edges are arguably as consistent through the census years. These spatial patterns along with the decreasing percentage of inadequate dwellings in the city, and the R^2, propose a long-term development, concentration, and strengthening of the relationship between dwellings in need of major repairs and the predictors identified in this research. This continued polarization of dwelling conditions in a number of areas in the city suggest that inadequate dwellings have been in place for many communities for years—a polarization that has contributed to a greater narrative of wealth and quality of life in Toronto (Hulchanski 2010).

By census tract, these communities have experienced high levels of inadequate dwellings, and based on the eight predictors, there are statistical and spatial results supporting this outcome. Figure 6.10 compares the 2016 census results of dwellings in need of major repairs to the top three predictors of the 2016 regression analysis by Pearson correlation result. The third highest Pearson correlation of "apartment in a building that has five or more floors" shows similar spatial patterns to inadequate dwellings, with the minor exception of the downtown on Yonge Street and at the waterfront. "Dwellings constructed over 20 or 25 years ago" and "household income

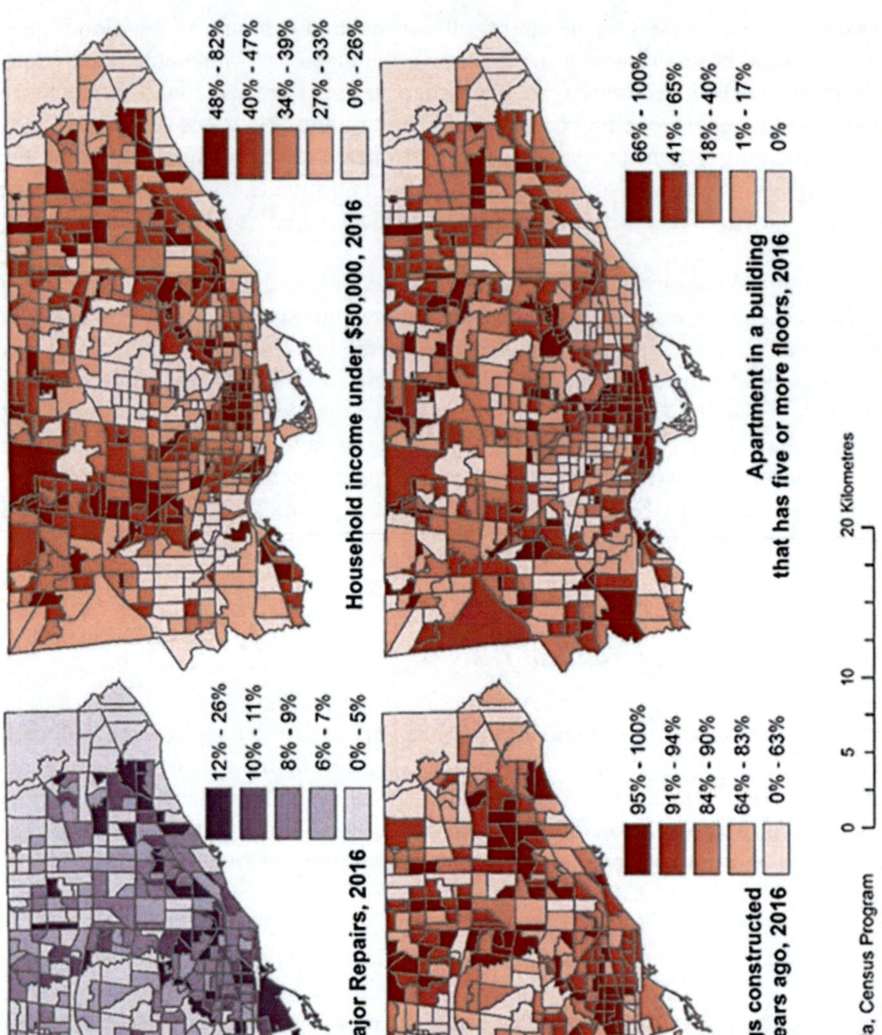

Fig. 6.10 2016—inadequate dwellings and the top three predictors by Pearson correlation, by census tract in the City of Toronto

under $50,000" show spatial patterns that are much more similar to the inadequate dwelling map. This would likely contribute to the higher Pearson correlation results as well as the argument that marginalized and vulnerable communities are more likely to be found in inadequate dwellings.

6.5.4 Limitations

The research takes a look a back at the past 20 years by census year due to census variable availability. The initial 18 predictors or potential independent census variables had to be present or capable of being aggregated and derived to achieve comparability and predictor consistency. The result of this decision was a study that only went back 20 years or 5 census years. One of the potential predictors was the census variable regarding visible minority populations, a variable which was only found in the census years from 1996 to the most recent census. Census variable instances like this were the reason behind the study only looking back to 1996.

Statistics Canada divides dwelling condition into homes in need of major repairs, minor repairs, and only regular maintenance needed. The last two were originally aggregated to a single variable, to act as a second dependent variable in the study. However, in the course of developing the explanatory model through the stepwise regression analyses by census year, it was discovered that consistently over the years a violation of independence was present. Specifically the VIF exhibited extremely high values over 30 and had selected both "dwellings constructed in the past 20 or 25 years" and "dwellings constructed over 20 or 25 years ago." This could be explained by the consistency over the census years, that over 90% of homes in Toronto only needed regular maintenance or minor repairs, and the stronger focus on predictors associated with marginalized and vulnerable groups from the literature.

The dwelling age proxy variable was also not divided in matter that was ideal for census year-to-year comparison. The inconsistency in year range resulted in aggregations that had census years 2001 and 2011 grouping older and younger dwellings between the twenty-year mark, while census years 1996, 2006, and 2016 grouped older and younger dwellings between the twenty-five year mark.

Finally, the slight dip of the R^2 trend and the stabilization of the percentage of inadequate dwelling could be explained by the change in census collection in the 2011 census year. The removal of the long-form census as compulsory at the time which had since returned for the 2016 census had brought concerns of relevance, use in policy, planning, risk of bias, and use in comparative research to the forefront (Baskerville 2010). These concerns may have impacted the quality of the 2011 census and therefore may be a factor in the drop in R^2 due to the ruling federal government's choices of the time. Future work will hopefully have the continued benefit of a compulsory long-form census from Statistics Canada, which would continue a level of confidence with the data for comparative research.

6.6 Conclusion and Progress

This research was possible through the census collection of Statistics Canada and had a focus on dwellings in need of major repairs, an indicator of inadequate dwellings by organizations concerned with housing (Statistics Canada 2017). The desire to contribute to the conversation of inadequate dwellings was based on the importance of housing, and how poor living conditions can be detrimental to the quality of life, especially for tenants (United Way 2011). Nationally, efforts from the federal government of Canada have already improved work toward dwelling repairs since September 2017, including $83.5 million from the "Renovation and Retrofit of Existing Federally Administered Community Housing" investment and $490.4 million from the "Supporting Energy- and Water-Efficiency Retrofits to Existing Community Housing" investment (Government of Canada 2017). These nationwide investments have targeted areas in need and have included Canada's largest city, the cultural and economic center of Toronto.

Based on the explanatory model of this research, the application of similar integrated methods for other regions throughout the world would be seen as beneficial. Numerous statistical analyses which build on or are complementary to each other in issues such as housing would be useful in other major Canadian cities. Internationally, the use of an integrated method in research over a number of years would be dependent on data quality and if comparable years of data exist. Both developed and undeveloped nations with open data on demographics could use integrated methods to explore the relationships within societal issues such as housing in their major cities.

Although this study was conducted with ArcGIS and SPSS, this important research could be replicated and continued using open-sourced software and applications such as QGIS, "R" which is supported by the R Foundation for Statistical Computing, and GeoDa. This spatial statistical study could be pushed in the future into local modeling. Geographically weighted regression (GWR) is an approach that estimates local regression coefficients which has a moving weighting kernel and engages all data weighted by a geographical kernel to interpret local variables of parameters (Nakaya 2001). This would improve the model at the census tract level.

By identifying the suggested relationships and circumstances regarding inadequate dwellings in Toronto, plans for repair and intervention can be built on this progress with more understanding. The research had identified trends in dwellings in need of major repairs, which included a decrease and concentration of inadequate dwellings in the city, a strengthening relationship with eight home and resident predictors, suggesting an increased association with marginalized and vulnerable communities, and spatial polarization in the city visualizing adequate and inadequate areas by census tract. Understanding the homes and residents who are more likely to be affected by inadequate dwellings and where they are will be critical in reaching out to communities and improving living conditions and interventions such as a National Housing Strategy.

Findings in this research can help decision-makers of the strategy when distributing $4.3 billion from the Canada Community Housing Initiative and $15.9 billion from the National Housing Co-Investment Fund with the goal of repairing 240,000 units across Canada (Government of Canada 2017). Together with insight from this research, these investments would make an impact on Canada's largest city. Equipped with a better understanding of the relationships and circumstances involved, our communities and our government can confront inadequate dwellings in the City of Toronto.

Appendix (Tables 6.10, 6.11 and 6.12)

Table 6.10 Census variable directory

Variable code	Full variable description
	Home characteristics
RDw_AgeU_25	Dwellings constructed in the past 20 or 25 years[a]
RDw_AgeO_25	Dwellings constructed over 20 or 25 years ago[a]
AvgNRoom	Average number of rooms per dwelling
R_Apartm	Apartment in a building that has five or more floors
Sh30morP	Tenant households spending 30% or more of its income on shelter costs
	Resident characteristics
@0to14	Residents aged 0 to 14[a]
@65plus	Residents aged 65 and over[a]
UNemploy2	Unemployment rate
Move1YR	Mobility—Moved in the past year
Move5YR	Mobility—Moved in the past 5 years
R3undOc	Household occupied by three or less residents[a]
R4upOc	Household occupied by four or more residents[a]
R1_VisM	Visible minority population
ImmigP	Immigrant population
Married	Married or living common law[a]
NotMarry	Not married and not living common law[a]
NonEngSpo	Language spoken most often at home—Nonofficial languages
U50Inc	Household income under $50,000[a]

[a]A derived variable aggregated from a number of variables

Table 6.11 Hierarchical cluster analysis—proximity matrix, part 1

Case	Matrix file input							
	RDw_AgeU_25 (20)	RDw_AgeO_25 (20)	AvgNRoom	R_Apartm	R1_VisM	@0to14	@65plus	UNemploy2
RDw_AgeU_25 (20)	0.000	0.986	0.085	0.994	0.996	0.996	0.996	0.995
RDw_AgeO_25 (20)	0.986	0.000	0.164	0.991	0.993	0.991	0.991	0.989
AvgNRoom	0.085	0.164	0.000	0.028	0.118	0.028	0.024	0.000
R_Apartm	0.994	0.991	0.028	0.000	0.995	0.997	0.998	0.997
R1_VisM	0.996	0.993	0.118	0.995	0.000	0.996	0.995	0.994
@0to14	0.996	0.991	0.028	0.997	0.996	0.000	1.000	1.000
@65plus	0.996	0.991	0.024	0.998	0.995	1.000	0.000	1.000
UNemploy2	0.995	0.989	0.000	0.997	0.994	1.000	1.000	0.000
Move1YR	0.996	0.990	0.012	0.998	0.995	1.000	1.000	1.000
Move5YR	0.997	0.994	0.084	0.998	0.998	0.999	0.998	0.998
U50Inc	0.996	0.995	0.060	0.999	0.997	0.999	0.999	0.998
Married	0.997	0.996	0.148	0.994	0.998	0.996	0.996	0.994
NotMarry	0.996	0.997	0.107	0.997	0.998	0.998	0.998	0.997
ImmigP	0.997	0.995	0.110	0.997	1.000	0.997	0.997	0.996
NonEngSpo	0.997	0.992	0.047	0.998	0.998	0.999	0.999	0.999
Sh30morP	0.997	0.996	0.111	0.997	0.998	0.998	0.997	0.996
R3undOc	0.993	0.998	0.171	0.993	0.996	0.992	0.993	0.990
R4upOc	0.998	0.993	0.082	0.996	0.998	0.999	0.999	0.998

Table 6.12 Hierarchical cluster analysis—proximity matrix, part 2

Case	Matrix File Input									
	Move1YR	Move5YR	U50Inc	Married	NotMarry	ImmigP	NonEngSpo	Sh30morP	R3undOc	R4upOc
RDw_AgeU_25 (20)	0.996	0.997	0.996	0.997	0.996	0.997	0.997	0.997	0.993	0.998
RDw_AgeO_25 (20)	0.990	0.994	0.995	0.996	0.997	0.995	0.992	0.996	0.998	0.993
AvgNRoom	0.012	0.084	0.060	0.148	0.107	0.110	0.047	0.111	0.171	0.082
R_Apartm	0.998	0.998	0.999	0.994	0.997	0.997	0.998	0.997	0.993	0.996
R1_VisM	0.995	0.998	0.997	0.998	0.998	1.000	0.998	0.998	0.996	0.998
@0to14	1.000	0.999	0.999	0.996	0.998	0.997	0.999	0.998	0.992	0.999
@65plus	1.000	0.998	0.999	0.996	0.998	0.997	0.999	0.997	0.993	0.999
UNemploy2	1.000	0.998	0.998	0.994	0.997	0.996	0.999	0.996	0.990	0.998
Move1YR	0.000	0.998	0.999	0.995	0.997	0.997	0.999	0.997	0.992	0.998
Move5YR	0.998	0.000	0.999	0.999	1.000	0.999	0.999	0.999	0.997	0.999
U50Inc	0.999	0.999	0.000	0.997	0.999	0.999	0.999	0.999	0.996	0.998
Married	0.995	0.999	0.997	0.000	0.999	0.999	0.997	0.999	0.999	0.998
NotMarry	0.997	1.000	0.999	0.999	0.000	0.999	0.998	1.000	0.998	0.999
ImmigP	0.997	0.999	0.999	0.999	0.999	0.000	0.999	0.999	0.997	0.999
NonEngSpo	0.999	0.999	0.999	0.997	0.998	0.999	0.000	0.998	0.994	0.999
Sh30morP	0.997	0.999	0.999	0.999	1.000	0.999	0.998	0.000	0.998	0.999
R3undOc	0.992	0.997	0.996	0.999	0.998	0.997	0.994	0.998	0.000	0.995
R4upOc	0.998	0.999	0.998	0.998	0.999	0.999	0.999	0.999	0.995	0.000

References

Baculi E, Fast V, Rinner C (2017) The Geospatial contents of municipal and regional open data catalogs in Canada. J Urban Reg Inf Syst Assoc 28(1):39–48

Baker E, Lester LH, Bentley R, Beer A (2016a) Poor housing quality: prevalence and health effects. J Prev Interv Community 44(4):219–232

Baker MG, Zhang J, Blakely T, Crane J, Saville-Smith K, Howden-Chapman P (2016b) Collaborating with a social housing provider supports a large cohort study of the health effects of housing conditions. BMC Public Health 16:1–9

Baker E, Beer A, Lester L, Pevalin D, Whitehead C, Bentley R (2017) Is housing a health insult? Int J Environ Res Public Health 14(6):1–18

Baskerville J (2010) Canada census 2011: not scientifically valid? Significance 7(4):172–172

Bedggood R, Perenyi A, Meyer D, Farquharson K, Johansson C, Bedggood P, Milgate G (2017) The living conditions of aboriginal people in Victoria. Energy Procedia 121(Complete):278–283

Brisson D, Covert J (2015) Housing instability risk among subsidized housing recipients: characteristics associated with late or nonpayment of rent. Soc Work Res 39(2):119–128

Campagna G (2016) Linking crowding, housing inadequacy, and perceived housing stress. J Environ Psychol 45(Complete):252–266

Carter T (2005) The influence of immigration on global city housing markets: the Canadian perspective. Urban Policy Res 23(3):265–286

Chisholm E, O'Sullivan K (2017) Using twitter to explore (un)healthy housing: learning from the #Characterbuildings campaign in New Zealand. Int J Environ Res Public Health 14(11):1–14

Chisholm E, Howden-Chapman P, Fougere G (2017) Renting in New Zealand: perspectives from tenant advocates. Kōtuitui: New Zealand J Soc Sci Online 12(1):95–110

Choi NG (1999) Differences in the quality of housing units occupied by elderly men versus elderly women. J Hous Elder 13(1–2):93–113

Colburn G, Allen R (2018) Rent burden and the great recession in the USA. Urban Stud 55(1):226–243

Corman H, Curtis MA, Noonan K, Reichman NE (2016) Maternal depression as a risk factor for children's inadequate housing conditions. Soc Sci Med 149(Complete):76–83

Cronley C, Nordberg A, Murphy ER, Twis MK (2018) A community of singers: employing exploratory cluster analysis to reveal profiles of psychosocial characteristics among members of a street choir for homeless and formerly homeless individuals. J Soc Distress Homeless 27(1):64–74

Dahmann NM, Dennison B (2013) Organizing for healthy communities: a report from public housing in Los Angeles. Prog Community Health Partnersh 7(1):77–82

Díaz McConnell E (2017) Rented, crowded, and unaffordable? Social vulnerabilities and the accumulation of precarious housing conditions in Los Angeles. Hous Policy Debate 27(1):60–79

Dyb E (2016) Housing first or no housing? Housing and homelessness at the end of alcohol and drug treatment. Int J Drug Policy 36(Complete):76–84

Evans GW, Wells NM, Moch A (2003) Housing and mental health: a review of the evidence and a methodological and conceptual critique. J Soc Issues 59(3):475–500

Filali R (2012) Housing conditions in Tunisia: the quantity-quality mismatch. J Housing Built Environ 27(3):317–347

Filandri M, Olagnero M (2014) Housing inequality and social class in Europe. Hous Stud 29(7):977–993

Getis A (2007) Reflections on spatial autocorrelation. Reg Sci Urban Econ 37(4):491–496

Ghosh S (2014) Everyday lives in vertical neighbourhoods: exploring Bangladeshi residential spaces in Toronto's inner suburbs. Int J Urban Reg Res 38(6):2008–2024

Gibler KM, Tyvimaa T, Kananen J (2014) The relationship between the determinants of rental housing satisfaction and considering moving in Finland. Prop Manag 32(2):104–124

Govender T, Barnes JM, Pieper CH (2011) Housing conditions, sanitation status and associated health risks in selected subsidized low-cost housing settlements in Cape Town. South Africa Habitat Int 35(2):335–342

Government of Canada (2017) Canada's national housing strategy: a place to call home. Government of Canada, Ottawa

Grösche P (2010) Housing, energy cost, and the poor: counteracting effects in Germany's housing allowance program. Energy Policy 38(1):93–98

Habib RR, Mahfoud Z, Fawaz M, Basma SH, Yeretzian JS (2009) Housing quality and ill health in a disadvantaged urban community. Public Health 123(2):174–181

Howden-Chapman P, Isaacs N, Crane J, Chapman R (1996) Housing and health: the relationship between research and policy. Int J Environ Health Res 6(3):173–185

Hulchanski JD (2010) The three cities within Toronto: income polarization among Toronto's neighbourhoods, 1970–2005. Cities Centre, University of Toronto, Toronto

Jacobs DE (2006) A qualitative review of housing hazard assessment protocols in the United States. Environ Res 102(1):13–21

Kohlhuber M, Mielck A, Weiland SK, Bolte G (2006) Social inequality in perceived environmental exposures in relation to housing conditions in Germany. Environ Res 101(2):246–255

Lee S, Parrott KR, Ahn M (2014) Housing adequacy: a Well-being indicator for elderly households in southern U.S. communities. Fam Consum Sci Res J 42(3):235–251

Lens M (2013) Subsidized housing and crime: theory, mechanisms, and evidence. J Plan Lit 28(4):352–363

Lin Y, Chang C, Chen C (2014) Why homebuyers have a high housing affordability problem: quantile regression analysis. Habitat Int 43(Complete):41–47

Liu Z, Huang J, Shi J, Tao R, Zhou W, Zhang L (2007) Characterizing and estimating rice brown spot disease severity using stepwise regression, principal component regression and partial least-square regression. J Zhejiang Univ Sci B 8(10):738–744

Milligan V, Pinnegar S (2010) The comeback of national housing policy in Australia: first reflections. Eur J Housing Policy 10(3):325–344

Canadian Mortgage and Housing Corporation (2014) Canadian housing observer: with a feature on housing affordability and need. Canadian Mortgage and Housing Corporation, Ottawa

Morton LW, Allen BL, Li T (2004) Rural housing adequacy and civic structure. Sociol Inq 74(4):464–491

Murdie RA (2003) Housing affordability and Toronto's rental market: perspectives from the housing careers of Jamaican, polish and Somali newcomers. Hous Theory Soc 20(4):183–196

Nakaya T (2001) Local spatial interaction modelling based on the geographically weighted regression approach. GeoJournal 53(4):347–358

Patterson ML, Somers JM, Moniruzzaman A (2012) Prolonged and persistent homelessness: multivariable analyses in a cohort experiencing current homelessness and mental illness in Vancouver, British Columbia. Ment Health Subst Use 5(2):85–101

Pendall R, Theodos B, Franks K (2012) Vulnerable people, precarious housing, and regional resilience: an exploratory analysis. Hous Policy Debate 22(2):271–296

Perkins DD, Larsen C, Brown BB (2009) Mapping urban revitalization: using GIS spatial analysis to evaluate a new housing policy. J Prev Interv Community 37(1):48–65

Pevalin DJ, Reeves A, Baker E, Bentley R (2017) The impact of persistent poor housing conditions on mental health: a longitudinal population-based study. Prev Med 105(Complete):304–310

Poortinga W, Jones N, Lannon S, Jenkins H (2017) Social and health outcomes following upgrades to a national housing standard: a multilevel analysis of a five-wave repeated cross-sectional survey. BMC Public Health 17:1–15

Purdy S (2003) "It was tough on everybody": low-income families and housing hardship in post-world war II Toronto. J Soc Hist 37(2):457–482

Reuschke D (2016) The importance of housing for self-employment. Econ Geogr 92(4):378–400

Rokach L, Maimon O (2005) Clustering methods. In: Data mining and knowledge discovery handbook. Springer, Boston, MA, pp 321–352

Rollings KA, Wells NM, Evans GW, Bednarz A, Yang Y (2017) Housing and neighborhood physical quality: Children's mental health and motivation. J Environ Psychol 50(Complete):17–23

Rosenbaum E, Friedman S (2004) Generational patterns in home ownership and housing quality among Racial/Ethnic groups in New York City, 1999. Int Migr Rev 38(4):1492–1533

Ross LM, Shlay AB, Picon MG (2012) You can't always get what you want: the role of public housing and vouchers in achieving residential satisfaction. City 14(1):35–53

Roy M, Génèreux M, Laverdière É, Vanasse A (2014) Surveillance of social and geographic inequalities in housing-related issues: the case of the eastern townships, Quebec (Canada). Int J Environ Res Public Health 11(5):4825–4844

Ruel E, Oakley D, Wilson G, Maddox R (2010) Is public housing the cause of poor health or a safety net for the unhealthy poor? J Urban Health 87(5):827–838

Schonlau M (2004) Visualizing non-hierarchical and hierarchical cluster analyses with clustergrams. Comput Stat 19(1):95–111

Shier M, Graham J, Fukuda E, Turner A (2016) Predictors of living in precarious housing among immigrants accessing housing support services. J Int Migr Integr 17(1):173–192

Spetic W, Kozak R, Cohen D (2008) How consumers value healthy houses: a preliminary segmentation of Canadian households. J Housing Built Environ 23(1):37–52

Statistics Canada (2017) Census of population reference guide: housing reference guide, census of population, 2016. Statistics Canada, Ottawa

Suzuki N, Ogawa T, Inaba N (2018) The right to adequate housing: evictions of the homeless and the elderly caused by the 2020 summer olympics in Tokyo. Leis Stud 37(1):89–96

Talen E, Koschinsky J (2014) The neighborhood quality of subsidized housing. J Am Plan Assoc 80(1):67–82

Tibesigwa BM, Hao L, Karumuna BV (2017) The concept of spatial quality and its challenges on exercised affordable housing design typology in Dar es Salaam – Tanzania. Habitat Int 59(Complete):44–59

Tucker M, Turley M, Holgate S (2014) Critical success factors of an effective repairs and maintenance service for social housing in the UK. Facilities 32(5/6):226–240

Tunstall R, Bevan M, Bradshaw J, Croucher K, Duffy S, Hunter C, Jones A, Rugg J, Wallace A, Wilcox S (2013) The links between housing and poverty: an evidence review. Joseph Rowntree Foundation, pp 4–84

UN Committee on Economic, Social and Cultural Rights (CESCR), General Comment No. 4: The Right to Adequate Housing (Art. 11 (1) of the Covenant), 13 (1991, December) E/1992/23, available at: http://www.refworld.org/docid/47a7079a1.html. Accessed 2 May 2018

van der Kloot WA, Spaans AMJ, Heiser WJ (2005) Instability of hierarchical cluster analysis due to input order of the data: the PermuCLUSTER solution. Psychol Methods 10(4):468–476

Vaz E, Arsanjani JJ (2015) Predicting urban growth of the greater Toronto area coupling a Markov cellular automata with document meta-analysis. J Environ Inform 25(2):71–80

Vaz E, Zhao Y, Cusimano M (2016) Urban habitats and the injury landscape. Habitat Int 56(Complete):52–62

Walters CR (2009) Do subsidized housing units depreciate faster than unsubsidized ones? J Hous Econ 18(1):49–58

Wang M, Wright J, Brownlee A, Buswell R (2016) A comparison of approaches to stepwise regression on variables sensitivities in building simulation and analysis. Energ Buildings 127(Complete):313–326

Wells NM, Harris JD (2007) Housing quality, psychological distress, and the mediating role of social withdrawal: a longitudinal study of low-income women. J Environ Psychol 27(1):69–78

Wen Z, Zhou MX (2008) Evaluating the use of data transformation for information visualization. IEEE Trans Vis Comput Graph 14(6):1309–1316

Windle GS, Burholt V, Edwards RT (2006) Housing related difficulties, housing tenure and variations in health status: evidence from older people in Wales. Health Place 12(3):267–278

Wright PA, Kloos B (2007) Housing environment and mental health outcomes: a levels of analysis perspective. J Environ Psychol 27(1):79–89
Yoon J (2015) Geographical variations and genetic distances of three saxidomus purpuratus populations ascertained by PCR analysis. Dev Reprod 19(4):259–264
Zimisuhara B, Valdiani A, Shaharuddin NA, Qamaruzzaman F, Maziah M (2015) Structure and principal components analyses reveal an intervarietal fusion in Malaysian mistletoe fig (ficus deltoidea jack) populations. Int J Mol Sci 16(7):14369–14394

Chapter 7
Ontario Wetland Policy Analysis

7.1 Introduction

The ecological and economic benefits of wetlands have been well documented in academic literature; wetlands provide sediment and nutrient retention, water filtration, biochemical transformation, water storage during flooding events, carbon sequestration, and a habitat for many native species (Bradford 2016; Yang et al. 2016). An additional important aspect of wetlands is their ability to hold organic carbon stock, and in Southern Ontario, the levels of peat and organic carbon stock have been traditionally underestimated (Byun et al. 2018). Wetlands have been negatively impacted in Southern Ontario since the arrival of European settlers in the 1700s, and the primary stressors to wetlands in this region are urban development and agricultural expansion (Croft-White et al. 2017). Agricultural expansion which allows for wetland drainage is considered the leading driver of wetland loss in Ontario (Walters and Shrubsole 2003).

A central issue with the management of wetlands in Ontario is that existing policies, although apparently robust, have several gaps which pose challenges for effective wetland conservation. The Ontario provincial government outlines the policy instruments that have been implemented to restrict certain activities in wetlands (i.e., to foster wetland protection). These policy instruments include "provincial instruments that restrict certain activities in wetlands and provincial instruments that facilitate wetland conservation" (Government of Ontario 2020a, b). However, the Environmental Commissioner of Ontario has stated in the 2018 report that current approaches to wetland management are not working (Environmental Commissioner of Ontario 2018).

There are several issues with the provincial management of wetlands including clarity regarding how provincial authorities should make decisions, lack of informed decision-making, enforcement, fragmented jurisdiction, communication between agencies, and determining the value of individual wetlands (i.e., only the most

E. Vaz, *Regional and Urban Change and Geographical Information Systems and Science*, Advances in Geographic Information Science,
https://doi.org/10.1007/978-3-031-24731-6_7

"valued" wetlands are afforded protection) (Schulte-Hostedde et al. 2007). Additionally, there are issues with lack of adherence and lack of enforcement by conservation authorities to the legislation outlined in the Drainage Act (Walters and Shrubsole 2003). The Ontario provincial government has also developed three major action items that begin to address some of the shortcomings mentioned above; however, there are some notable flaws to these major actions (Government of Ontario 2020a, b; Environmental Commissioner of Ontario 2018).

This paper argues that there are several shortcomings to the current policies used to facilitate wetland management, and improvement in these policies is needed to ensure effective wetland conservation. This paper will discuss the following: the context of this issue, a review of current wetland management approaches implemented by the Ontario provincial government, a critique of the current wetland management approaches, alternative approaches implemented in other North American regions and considerations for implementing these approaches in Ontario, and conclusions.

It is noted that this paper and the literature used to support this critique focus on wetland loss in Southern Ontario because (1) this is the most highly populated region in the province and (2) has experienced and continues to experience the most dramatic wetland loss compared to Northern Ontario (Byun et al. 2018).

Wetland management is complex and involves cooperation between the federal government, provincial government, municipal governments, conservation authorities, and landowners (Government of Ontario 2020a, b). This paper focuses on three central problems with the management of wetland conservation in Ontario: the general shortcomings of the provincial management, the limitations of the Drainage Act and its lack of enforcement by conservation authorities, and the flaws of the major actions to be implemented by the provincial government. Discussions included below will highlight limitations to the Drainage Act, the lack of adherence to the guidelines in the Drainage Act, and the lack of enforcement to these guidelines by conservation authorities. Additionally, to add to the discussion of the complexity of wetland management and conservation, this paper also discusses more generally the provincial wetland management and the shortcomings of existing policies. The more general shortcomings of existing policies are to varying degrees connected to the more pressing issue of wetland drainage.

7.2 Context

Wetlands are both economically and ecologically important, and they provide a habitat for many ecologically important species, prevent erosion, reduce the impacts of flooding events, maintain water tables, and facilitate water filtration (Greb et al. 2006). In the province of Ontario, the economic benefits that wetlands provide are estimated to be at least $14 billion, and they provide a reduction in costs associated with flooding events by approximately 38% (Troy and Bagstad 2013; Moudrak, et al. 2017). In Southern Ontario, wetland loss is estimated to be at 68% (since the arrival of European settlers) (Bradford 2016). There are several anthropogenic drivers of wetland loss including agricultural drainage, conversion for fish production and logging, construction (e.g., dams and canals), peat extraction, and urban and suburban development (van Asselen et al. 2013). However, wetland drainage for the purpose of agricultural expansion is the leading driver of wetland loss in the province and contributes to an estimated 81–85% of wetland loss (Walters and Shrubsole 2003).

Wetland drainage is an important aspect of agriculture, with the intention of improving the productivity of the land (Walters and Shrubsole 2005). Wetland drainage is done to increase crop production, and this is accomplished by creating an environment where agricultural vegetation is better able to increase nutrient uptake (Zucker and Brown 1998). However, wetland drainage for agricultural expansion has several direct and indirect environmental consequences including habitat loss, water quality reduction, hydrologic alteration, and increases in sediment, phosphorus, and nitrogen runoff (Blann et al. 2009).

7.3 Current Approaches to Wetland Management in Ontario

There are several policies in the province of Ontario that aid in wetland management including 20 pieces of legislation, which are implemented by a provincial agency, federal departments, and provincial ministries, 36 conservation authorities, and 444 different municipalities (Government of Ontario 2020a, b). This section outlines the existing provincial wetland conservation Acts and policy instruments, the role that conservation authorities play in wetland conservation under the Conservation Authority Act and the Drainage Act, and the wetland major actions to be implemented by the provincial government.

7.3.1 Provincial Wetland Conservation Acts and Policy Instruments

The policy instruments that currently aid in the protection of wetlands include the Planning Act, the Niagara Escarpment Planning and Development Act and Plan, the Oak Ridges Moraine Conservation Act, the Greenbelt Act, the Places to Grow Act, the Lake Simcoe Protection Act, the Conservation Authorities Act, the Environmental Protection Act, the Crown Forest Sustainability Act, the Public Lands Act, the Lakes and Rivers Improvement Act, and the Water Resources Act (Government of Ontario 2020a, b).

In addition to the Acts, mentioned above, that aid in the protection and conservation of wetlands is the Drainage Act, which regulates wetland drainage in the province of Ontario. Within this Act, "municipalities are legislated to maintain and repair drains and to respond to petitions for new drainage systems [and] under Section 28 of the *CA Act*, conservation authorities (CAs) regulate development in or adjacent to watercourses, wetlands, the shoreline of the Great Lakes-St. Lawrence River System or inland lakes, river or stream valleys, hazardous lands and other areas" (Government of Ontario 2020a, b). This Act, coupled with various statutes, provides the provincial government statutory authority over private property (in regard to drainage) and authority over public lands and natural resources (Walters and Shrubsole 2005).

7.3.2 Ontario Conservation Authorities: The Conservation Authority Act and the Drainage Act

As mentioned previously, there are several Acts and policy instruments that have been implemented by the provincial government to aid in wetland conservation. Outlined in the Conservation Authority Act (and supported through the Drainage Act) is a set of regulations which give power to conservation authorities throughout the province to aid in wetland conservation. This Act was amended in 1998 to establish consistency between different conservation authorities throughout the province, and this amendment ensured that all conservation authorities were regulating all water bodies including wetlands. The role that conservation authorities play in wetland conservation is (1) ensuring that the policies under the Policy Act are upheld; (2) ensuring permission is granted to landowners before an area is developed if development may lead to erosion, pollution, dynamic beaches, or flooding; and (3) to regulate any changes, diversion, straightening, or interfering development that may take place on any water body including wetlands (Conservation Ontario 2020).

7.3.3 Wetland Conservation Major Actions

Included in the Ontario provincial government's Wetland Conservation Strategy are three major actions aimed at preventing the further degradation of wetlands in the province, and they are outlined below.

1. Improving Ontario's Wetland Inventory and Mapping: This action aims to pair wetland location area and quality data with trend analysis to provide a more accurate understanding of the location and extent of wetlands throughout the province. This information will be used to measure and improve land-use protocols and policies. The Ontario provincial government recognizes that there is room for improvement in their current wetland map inventory and notes that mapping techniques need to be updated and standardized and incorporate new information like climate change scenarios (Government of Ontario 2020a, b).

2. Creating a No Net Loss Policy for Ontario's Wetlands: The purpose of this action is to implement a wetland offsetting policy which will allow for the creation of new artificial wetlands to replace wetlands lost to development and drainage. This action recognizes that due to the sensitivity of many ecosystems, many wetlands will be ineligible for the offsetting policy. Additionally, important considerations of this policy include recognition that offset wetlands must replicate an equal than or greater area and function of the original wetland, provincial oversight must be provided, and only certain types of land use and resource extraction would qualify for wetland offsetting (Government of Ontario 2020a, b).
3. Improving Guidance for the Evaluation of Significant Wetlands: This action will continue to use the Ontario Wetland Evaluation System (OWES) created in the 1980s to identify provincially significant wetlands. This system uses a numerical ranking system to rank wetlands based on their natural processes and their societal value. The continued use of this system allows for the merging of new and existing information on wetlands in the province and for the province to use this information for informed decision-making (Government of Ontario 2020a, b).

7.4 Critique of Current Approaches

Although it appears that wetland management is given adequate attention and resources by the Ontario government, there are still several shortcomings of the provincial management of wetlands. This section provides a critique of the provincial management of wetlands, the Drainage Act and limitations faced by conservation authorities, and limitations to the major actions that the provincial government is currently working.

7.4.1 Critique of Provincial Management

As mentioned previously, based on the number of Acts and policy instruments implemented by the Ontario provincial government, it appears that there is sufficient action being taken by the province to ensure wetland conservation is adequately managed in Ontario. However, there are several shortcomings in the provincial management of wetlands including lack of informed decision-making, determining the value of individual wetlands (i.e., only the most "valued" wetlands are afforded protection), fragmented jurisdiction, clarity regarding how provincial authorities should make decisions, enforcement, and communication between agencies (Schulte-Hostedde et al. 2007).

The shortcomings outlined by Schulte-Hostedde et al. (2007) are echoed by the 2018 Environmental Commissioner of Ontario report. This report emphasizes that despite current efforts from the provincial government, wetlands in Ontario continue to be destroyed (Environmental Commissioner of Ontario 2018).

Regarding lack of informed decision-making, a major shortcoming highlighted in the 2018 report is that the Ontario Ministry of Agriculture, Food and Rural Affairs (OMAFRA) fails to actually track the impacts of agricultural drainage on wetlands and cannot provide sufficient data on the extent and number of wetlands lost to agricultural drainage in the province. Additionally, the location and extent of peat and topsoil extraction (which represent about 7% of wetland loss) are also not tracked in the province, and there is actually no existing province-wide policy that prohibits peat and topsoil extraction (Environmental Commissioner of Ontario 2018).

Regarding the value of individual wetlands, the 2018 report highlights a flaw that is important to acknowledge: "legal protection to wetlands applied only to wetlands that have been identified as 'significant'" (Environmental Commissioner of Ontario 2018). It is required that municipalities map all provincially significant wetlands (PSWs) and prohibit destruction of PSWs before applications for development or drainage are accepted. However, the central problem with this system is that this lengthy, labor-intensive, and expensive evaluation process leaves many wetlands unprotected. Additionally, wetlands are deemed either provincially significant or not provincially significant, meaning that wetlands that score just below the significance threshold are not afforded any greater protection than those which score well below the significance threshold (Environmental Commissioner of Ontario 2018). In Ontario, wetlands are evaluated through the Ontario Wetlands Evaluation System (OWES) which scores wetlands on over 50 variables to determine their functional significance and maps wetland extent through aerial photographs, as well as field and mapping analysis. This process, in addition to being lengthy, labor intensive, and expensive, often fails to account for small wetlands (less than 0.5 ha) because they are more challenging to map (Environmental Commissioner of Ontario 2018). The 2018 report discusses that a possible solution to this particular shortcoming is to flip this policy (or reverse the onus), meaning that all wetlands would be afforded the title of "significant" until proven otherwise (Environmental Commissioner of Ontario 2018).

Clarity regarding how provincial authorities should make decisions and jurisdictional fragmentation is another issue with wetland management that requires improvement. The 2018 report discusses issues with the Provincial Policy Statement (PPS) and highlights that it provides only limited protection and only addresses the issue of development on wetland loss. The lack of information in the PPS has led to a disparity in how conservation authorities and land-use planners interpret the PPS guidelines. It is noted that the PPS has unclear wording which has led to differentiation in how jurisdictions interpret existing versus new agricultural land, and what constitutes "development." Additionally, the PPS guidelines only apply to PSWs, meaning that wetlands scoring below the significance threshold are not afforded any protection under the PPS (Environmental Commissioner of Ontario 2018). This problem is exacerbated by the lack of available information and challenges with the current system used to identify PSWs discussed previously (Environmental Commissioner of Ontario 2018).

The lack of enforcement and communication between different agencies is highly connected to limitations faced by conservation authorities throughout the province. Even though the Conservation Authorities Act was amended in 1998 to establish consistency between different conservation authorities throughout the province, inconsistencies and conflicting priorities remain (Conservation Ontario 2020; Environmental Commissioner of Ontario 2018). These problems, highlighted in the 2018 report, include lack of consistency with how conservation authorities exercise their powers, conflicting priorities between conservation authorities, inconsistent resource restraints, vague language used to guide conservation authorities, and an overall lack of guidance from the provincial government (Environmental Commissioner of Ontario 2018). The Conservation Authorities Act, used to guide how conservation authorities enforce wetland conservation, contains vague language that discourages the enforcement of wetland protection and creates challenges for conservation authorities in determining the extent to which wetland drainage and filling activities may be regulated or enforced. Additionally, the language used in the Conservation Authorities Act creates challenges in conservation authorities determining if any interference with a wetland has actually occurred. Requests from conservation authorities for the provincial government to provide an adequate list or definition of activities that interfere with wetlands, as well as an adequate definition of what constitutes a wetland, have not been granted (Environmental Commissioner of Ontario 2018). These challenges faced by conservation authorities mean that they are unable to "regulate threats to all wetlands and all threats to wetlands" (Environmental Commissioner of Ontario 2018).

Overall this report highlights that current approaches to wetland management are not working; agricultural drainage continues to be the leading cause of wetland depletion in the province; and there is a great deal of variance with how conservation authorities regulate wetlands, and a struggle faced by conservation authorities is the lack of provincial direction (Environmental Commissioner of Ontario 2018).

7.4.2 *Critique of the Drainage Act: Limitations Faced by Conservation Authorities*

As mentioned previously, wetland drainage is regulated by the Ontario provincial government under the Drainage Act. In order for landowners to acquire permission to drain water (e.g., a wetland) from their land, the Drainage Act requires that they follow the 38 steps outlined in the drainage legislation (including providing signed petitions, gaining approval from their local council, resolving all appeals, and gaining approval from the Ministry of Agriculture, Food and Rural Affairs) and continue to follow the maintenance, repair and improvement, and abandonment of a drain or part of drain requirements (Ministry of Agriculture, Food and Rural Affairs 2020).

However, a central shortcoming of the Drainage Act is there are no direct penalties or incentives in place that actually prevent agricultural wetland drainage

(Walters and Shrubsole 2003). This is a central problem with managing the impacts of agricultural drainage on wetland loss; there is an expectation that farmers obtain permission to drain water from their land; however, in practice this does not always happen (Cortus et al. 2009).

The current legislation outlined in the Drainage Act coupled with the shortcomings of the provincial government and limitations faced by conservation authorities throughout the province show that the current system for regulating wetland drainage is not working. Landowners are expected to adhere to the guidelines outlined in the Drainage Act; however, this does not always happen, and conservation authorities face many limitations (discussed previously) which add additional challenges to enforcing these regulations (Ministry of Agriculture, Food and Rural Affairs 2020; Cortus et al. 2009; Environmental Commissioner of Ontario 2018). Subsequently, the Environmental Commissioner of Ontario report argues that private landowners must be encouraged to implement wetland stewardship on their land (Environmental Commissioner of Ontario 2018).

7.4.3 Critique of Major Actions

Outlined previously were the major actions aimed at preventing the further degradation of wetlands in the province, which included improving Ontario's wetland inventory and mapping, creating no net loss policy for wetlands in Ontario, and improving guidance for the evaluation of significant wetlands. Although these actions may improve the management of wetlands in Ontario, there are some issues that are important to acknowledge (Government of Ontario 2020a, b).

First, regarding improvements to Ontario's wetland inventory and mapping, there is not any noteworthy criticism to be discussed. The limitations of mapping wetlands are highlighted in the Environmental Commissioner of Ontario's 2018 report, which discusses that mapping and creating an inventory of wetlands is a lengthy, labor-intensive, and expensive process which often fails to account for small wetlands (Environmental Commissioner of Ontario 2018). These limitations are acknowledged by Ontario's wetland inventory and mapping action item (Government of Ontario 2020a, b).

Second, regarding the creation of a no net loss policy for wetlands in Ontario, there are some limitations to this process that are important to acknowledge. The Nature Conservancy of Canada highlights problems with no net loss strategies and argues that these strategies rarely replicate the same features and functions that were initially in place. They often fail to fully replicate the complexity of the ecology or ecosystems lost, the time between wetland loss and new wetland formation has a lag time which often results in net biodiversity loss, and they often fail to prevent the loss of species at risk (Nature Conservancy of Canada 2020).

Third, regarding the improvement of guidance for the evaluation of significant wetlands, this major action outlined by the provincial government fails to account for the Environmental Commissioner of Ontario's report recommendation. As stated previously, the 2018 report advises to "reverse the onus" or, in other words, allow all wetlands to be afforded the title of "significant" until proven otherwise (Environmental Commissioner of Ontario 2018). However, the improvement of guidance for the evaluation of significant wetlands' major action calls for the continued use of the Ontario Wetland Evaluation System (OWES) (Government of Ontario 2020a, b). As mentioned previously, this business as usual use of the OWES will continue to jeopardize the conservation of many wetlands throughout the province (Environmental Commissioner of Ontario 2018).

7.5 Alternative Approaches

This paper has outlined several problems with the management of wetlands in Ontario including shortcomings of the provincial government (i.e., lack of informed decision-making, determining the value of individual wetlands, fragmented jurisdiction, clarity regarding how provincial authorities should make decisions, enforcement, and communication between agencies) (Schulte-Hostedde et al. 2007), shortcomings of the Drainage Act and limitations faced by conservation authorities (i.e., lack of adherence from landowners and lack of enforcement by conservation authorities) (Cortus et al. 2009; Environmental Commissioner of Ontario 2018), and shortcomings of major actions to be implemented by the provincial government (i.e., improving Ontario's wetland inventory and mapping, creating no net loss policy for wetlands, and improving guidance for the evaluation of significant wetlands) (Government of Ontario 2020a, b).

When critiquing the shortcomings of provincial wetland management, it should be noted that balancing the management of wetland with economic, societal, and other environmental concerns is highly a complex undertaking. It is understandable that a perfect system for wetland management is not in place in Ontario and shortcomings of the provincial management of wetlands are expected. Rather than providing recommendations that aim to resolve all of the criticisms discussed, this section aims to discuss alternative approaches to wetland management that have been used in other regions that may be applied to the province of Ontario.

As discussed previously, a major shortcoming to the Drainage Act is the lack of adherence from landowners and the lack of enforcement to this legislation. A possible solution to this problem, which has been implemented in the United States, is the introduction of voluntary programs which give landowners incentives to restore or conserve wetlands. The Wetlands Reserve Program in particular has focused on optimizing wetland functions and values as well as wildlife habitat in every agricultural acreage enrolled in their program. Programs like this have been highly successful in restoring wetlands and increasing wildlife habitat (Benson et al. 2018). An additional successful strategy of wetland restoration and conservation exists in Connecticut in the United States. In this state, there is a bottom-up approach to the management of wetlands where the Connecticut Inland Wetlands and Watercourses Act provides regulatory guidance but has given jurisdiction of land-use activities surrounding wetlands to local towns and municipalities. Within this system, impacts from agricultural activities on wetlands have actually been very small, and the natural resources have been managed in an environmentally sustainable manner (Owens and Zimmerman 2013). In regard to implementing voluntary programs such as these in Ontario, it would be important to provide landowners with enough information to understand the importance of wetland conservation. A study conducted by Nebel and others (2017) sought to better understand which factors contribute to environmental behavior changes in landowners in southwestern Ontario. When looking at landowners' likelihood of entering a wetland enhancement program to either restore or conserve wetlands on their property, this study found that motivation to enroll was enhanced when "access to more information on how the decline in wetlands area affects them personally [compared to] public recognition as the lest motivating factor" (Nebel et al. 2017). Additionally, this study found that monetary incentives were less likely to motivate landowners to enter a wetland enhancement program and likelihood of enrollment was higher if the landowners relied on their land as their primary source of income (Nebel et al. 2017).

Another example of an area which has experienced historical massive wetland loss which for the past 30 years has adopted more environmentally conscious and sustainable wetland management is the Florida Everglades. Drainage of the Florida Everglades has taken place as early as the 1880s with the state providing drainage contracts to several investors to facilitate economic expansion (Glenn 1999). However, in response to the increase of several environmental issues (e.g., water quality degradation) exacerbated by the drainage and therefore loss of the Florida Everglades, the use of adaptive management and adaptive governance has improved the overall functionality and increased the extent to this extensive wetland area (Gunderson and Light 2005). Adaptive governance acknowledges the complexity of managing a system (like wetlands) and integrates decision-making, science, and policy. This form of governance prepares for changes in management strategies and allows for the flexibility to deal with challenges that would traditionally impede adaptive management. Adaptive management is a form of resource management that uses scientific approaches and experiments to test policies and accounts for the dynamic nature of systems like wetlands. Although the Florida Everglades are far from being restored to their original extent, the implementation of adaptive management and adaptive governance strategies has significantly improved their functionality and extent (Gunderson and Light 2005).

In addition to these approaches to wetland management that have already been implemented and enhanced wetland conservation and restoration in their respective regions, there is an additional study that aims to highlight strategies to wetland management that may be implemented to improve wetland conservation. This study was conducted by Clare and others (2011), and it discusses the importance of implementing watershed-based planning, comprehensive economic and social valuation of wetlands, and long-term citizen-based monitoring schemes (Clare et al. 2011). Watershed-based planning uses a broad landscape context for wetland management which allows for large-scale systematic conservation planning (SCP), a framework that facilitates regional decision-making and landscape planning. SCP can be highly beneficial as it allows for a variety of perspectives to be included in the management process, and it allows for local and regional wetland management to work in tandem (Clare et al. 2011). Comprehensive economic and social valuation of wetlands allows for the true economic cost of wetland loss to be taken on by the proponent promoting the wetland loss, rather than the public. This process does so by "incorporating economic and social valuation processes into wetland permit approvals [that] may help link the desired ecosystem goods and services to benefit cost analyses of areas being considered for development" (Clare et al. 2011). Long-term citizen-based monitoring is a type of program that allows local citizens to aid in monitoring and managing local wetlands. This program would encourage citizen scientists to record and submit data on local wetlands and would provide a better understanding on the status and health of wetlands, and it also would also empower citizens to engage in local stewardship and put pressure on politicians to practice wetland conservation and restoration (Clare et al. 2011).

Strategies like implementing voluntary wetland conservation programs, adaptive management and adaptive governance, watershed-based planning, comprehensive economic and social valuation of wetlands, and long-term citizen-based monitoring schemes are all tools that could be used in Ontario to both improve the general provincial management of wetlands and address the issues of lack of adherence and lack of enforcement to the Drainage Act.

7.6 Conclusion

The economic and ecological importance of wetlands is well understood; however, development and agricultural expansion continue to facilitate wetland loss (Greb et al. 2006; Bradford 2016; van Asselen et al. 2013; Walters and Shrubsole 2003). Wetland drainage continues to be the leading driver of wetland loss in the province of Ontario, and there are several issues with approaches to wetland management that allow for this problem to continue (Walters and Shrubsole 2003; Environmental Commissioner of Ontario 2018; Cortus et al. 2009). This paper has argued that there are several shortcomings to the current policies used to facilitate wetland management, and improvement in these policies is needed to ensure effective wetland conservation.

The provincial management of wetlands includes several conservation Acts and policy instruments aimed at conserving and restoring wetlands, conservation authorities have been granted powers (through the Conservation Authority Act and the Drainage Act) to aid in wetland conservation, and the provincial government has proposed three major actions aimed at preventing further degradation of wetlands in the province (Government of Ontario 2020a, b; Conservation Ontario 2020). However, this paper has highlighted several shortcomings of the provincial management of wetlands, the limitations faced by conservation authorities, the lack of adherence of landowners to comply with the guidelines outlined in the Drainage

Act, and the flaws of the major actions to be implemented by the provincial government (Environmental Commissioner of Ontario 2018).

This paper has also discussed several strategies (implementing voluntary wetland conservation programs, adaptive management and adaptive governance, watershed-based planning, comprehensive economic and social valuation of wetlands, and long-term citizen-based monitoring schemes) that may aid in resolving some of the shortcomings discussed.

It is acknowledged that wetland management is highly complex and involves many actors. This paper has not aimed to provide solutions to all of the issues raised, but rather to highlight several strategies that either have been successfully implemented in other regions or in theory have the potential to be implemented in Ontario with the goal of improving wetland management in the province. The hope for this paper is that it provides a broad understanding of the central shortcomings of wetland management in Ontario. The economic and ecological significance of wetlands have historically been highly undervalued, and the current gaps in wetland management indicate that the importance of wetlands is still not fully appreciated.

Many of the issues discussed in this paper are not unique to Ontario, wetland loss is a global problem that has existed for over 100 years, and in more recent years, wetlands have experienced a greater loss (between 1950 and 2008) (Wójcicki and Woskowicz-Ślęzak 2015). Like in Ontario, the primary factors that contribute to wetland loss globally are agricultural expansion and urban development (Hu et al. 2017). Beyond highlighting the shortcomings that allow for continued wetland degradation in Ontario, additional goals of this paper are to (1) acknowledge that management issues seen in Ontario are likely seen in other governments around the world and (2) discuss strategies to aid in these issues that may also be applied to other governments around the world.

Wetlands are incredibly important at both local and global levels, and water purification, flood management, storm protection, and providing a habitat for many important species are just some of the functions that wetlands provide (Mao et al. 2018). Wetlands are considered multiple-value systems because of the multiple processes that they perform that are valued by humans (Mitsch and Gosselink 2000). It is highly important that the Ontario provincial government continues to implement policies that facilitate wetland conservation and restoration and also make improvements to existing management strategies that will allow for improvement in wetland conservation and ultimately allow for significantly fewer wetlands to be lost to anthropogenic activities.

References

Benson CE, Carberry B, Langen TA (2018) Public–private partnership wetland restoration programs benefit species of greatest conservation need and other wetland-associated wildlife. Wetl Ecol Manag 26:195–211. https://doi.org/10.1007/s11273-017-9565-8

Blann KL, Anderson JL, Sands GR, Vonderacek B (2009) Effects of agricultural drainage on aquatic ecosystems: a review. Crit Rev Environ Sci Technol 39(11):909–1001. https://doi.org/10.1080/10643380801977966

Bradford A (2016) Averting degradation of southern Ontario wetlands due to hydrologic alterations associated with development. Can Water Resour J 41(4):543–553. https://doi.org/10.1080/07011784.2015.1119061

Byun E, Finkelstein SA, Cowling SA, Badiou P (2018) Potential carbon loss associated with post-settlement wetland conversion in Southern Ontario, Canada. Carbon Balance Manag 13(6):1–12. https://doi.org/10.1186/s13021-018-0094-4

Clare S, Krogman N, Foote L, Lemphers N (2011) Where is the avoidance in the implementation of wetland law and policy? Wetl Ecol Manag 19:165–182. https://doi.org/10.1007/s11273-011-9209-3

Conservation Ontario (2020) Conservation Authorities Act. Retrieved from https://conservationontario.ca/conservation-authorities/planning-and-regulations/conservation-authorities-act-section-28-regulations/

Cortus BG, Unterschultz JR, Jeffery SR, Boxall PC (2009) The impacts of agriculture support programs on wetland retention on grain farms in the prairie pothole region. Can Water Resour J 34(3):245–254. https://doi.org/10.4296/cwrj3403245

Croft-White MV, Cvetkovic M, Rokitnicki-Wojcik D, Midwood JD, Grabas GP (2017) A shoreline divided: twelve-year water quality and land cover trends in Lake Ontario coastal wetlands. J Great Lakes Res 43:1005–1015. https://doi.org/10.1016/j.jglr.2017.08.003

Environmental Commissioner of Ontario (2018) Environmental protection report. Retrieved from https://docs.assets.eco.on.ca/reports/environmental-protection/2018/Back-to-Basics-Volume4-Ch1.pdf

Glenn J (1999) The origins of everglades drainage in the progressive era: local state and federal cooperation and conflict. ProQuest Dissertations Publishing, pp 1–152. Retrieved from https://search-proquest-com.ezproxy.lib.ryerson.ca/docview/304515343?pq-origsite=summon

Government of Ontario (2020a) Archived – wetland conservation strategy. Retrieved from https://www.ontario.ca/page/wetland-conservation-strategy

Government of Ontario (2020b) Drainage Act and conservation authorities act protocol. Retrieved from https://www.ontario.ca/page/drainage-act-and-conservation-authorities-act-protocol

Greb SF, DiMichele WA, Gastaldo RA (2006) Evolution and importance of wetiands in earth history. Geol Soc Am Spec Pap 399:1–40. https://doi.org/10.1130/2006.2399(01)

Gunderson L, Light SS (2005) Adaptive management and adaptive governance in the everglades ecosystem. Policy Sci 39(4):323–334. https://doi.org/10.1007/s11077-006-9027-2

Hu S, Niu Z, Chen Y, Li L, Zhang H (2017) Global wetlands: potential distribution, wetland loss, and status. Sci Total Environ 586:319–327. https://doi.org/10.1016/j.scitotenv.2017.02.001

Mao D, Wang Z, Wu J, Wu B, Zeng Y, Song K, Yi K, Luo L (2018) China's wetlands loss due to urban expansion. Land Degrad Dev 29(8):2644–2657. https://doi.org/10.1002/ldr.2939

Ministry of Agriculture, Food and Rural Affairs (2020) Drainage legislation. Retrieved from http://www.omafra.gov.on.ca/english/engineer/facts/89-166.htm

Mitsch WJ, Gosselink JG (2000) The value of wetlands: importance of scale and landscape setting. Ecol Econ 35(200):25–33. https://doi.org/10.1016/S0921-8009(00)00165-8

Moudrak N, Hutter AM, Feltmate B (2017) When the big storms hit: the role of wetlands to limit urban and rural flood damage. Ontario's Ministry of Natural Resources and Forestry. Intact Centre on Climate Adaptation. University of Waterloo

Nature Conservancy of Canada (2020) Why no net loss in biodiversity offsets fails nature and people. Retrieved from https://www.natureconservancy.ca/en/blog/archive/why-no-net-loss-in.html

Nebel S, Brick J, Lantz VA, Trenholm R (2017) Which factors contribute to environmental behaviour of landowners in Southwestern Ontario, Canada? Environ Manag 60:454–463. https://doi.org/10.1007/s00267-017-0849-9

Owens K, Zimmerman C (2013) Local governance versus centralization: Connecticut wetlands governance as a model. Rev Policy Res 30(6):629–656. https://doi.org/10.1111/ropr.12050

Schulte-Hostedde B, Walters D, Powell C, Schubsole D (2007) Wetland management: an analysis of past practice and recent policy changes in Ontario. J Environ Manag 82:83–94. https://doi.org/10.1016/j.jenvman.2005.12.007

Troy A, Bagstad K (2013) Estimation of ecosystem service values for Southern Ontario. Ontario Ministry of Natural Resources, Peterborough

van Asselen S, Verberg PH, Vermaat JE, Janse JH (2013) Drivers of wetland conversion: a global meta-analysis. PLoS One 8(11):1–13. https://doi.org/10.1371/journal.pone.0081292

Walters D, Shrubsole D (2003) Agricultural drainage and wetland Management in Ontario. J Environ Manag 69(4):369–379. https://doi.org/10.1016/j.jenvman.2003.09.013

Walters D, Shrubsole D (2005) Assessing efforts to mitigate the impacts of drainage on wetlands in Ontario, Canada. Can Geogr 49(2):155–171. https://doi.org/10.1111/j.0008-3658.2005.00086.x

Wójcicki K, Woskowicz-Ślęzak B (2015) Anthropogenic causes of wetland loss and degradation in the lower Kłodnica Valley (Southern Poland). Environ Soc Econ Stud 3(4):20–29. https://doi.org/10.1515/environ-2015-0070

Yang W, Liu Y, Ou C, Gabor S (2016) Examining water quality effects of riparian wetland loss and restoration scenarios in a Southern Ontario watershed. J Environ Manag 174:26–34. https://doi.org/10.1016/j.jenvman.2016.03.001

Zucker LA, Brown LC (1998) Agricultural drainage: water quality impacts and subsurface drainage studies in the Midwest. In: Bulletin. Minnesota University of Minnesota Extension, St. Paul, pp 871–898

Chapter 8
The Spatial and Temporal Distribution of Airbnb and Its Potential Impact on the Rental Market: A Case Study of City of Toronto, Ontario, Canada

8.1 Introduction

This chapter investigates connection between Airbnb rental proliferation and the price of apartment rentals in urban areas through a spatial and temporal analysis of Airbnb listing in the City of Toronto, Ontario, from 2016 to 2019. In doing so, it investigates both changes in locations of listings and the impacts these changes have had on long-term rental accommodation rates.

The digitization of businesses has changed the way we purchase goods and services and interact with one another. This has been amplified in the recent years, and especially in 2020 due to the COVID-19 pandemic. Digital platform economies such as Amazon, Uber, and—relevant to this research—Airbnb have experienced tremendous growth in the recent years. The short-term accommodation market, specifically Airbnb, has made a record-breaking $USD 101 billion initial public offering (IPO) even with the 85% decrease in booking rates due to the COVID-19 pandemic. The platform is widely used globally, with a total of 150 million worldwide users, as well as 5.6 million global Airbnb listings (Airbnb Statistics 2020).

In short, Airbnb puts those seeking short-term accommodation in touch with those who have housing that matches this demand using the Internet as the medium of transaction and communication. This service is undermining traditional markets for housing and tourism, mainly due its network of platforms that link individual demand to individual supply, with the company gaining financial profits. The private short-term accommodation market has raised many public policy questions, some of which are concerned with how these markets should be regulated and taxed. Others question the impact of these activities on housing and rental markets.

Short-term rentals are not a new phenomenon, but the availability of online platforms such as Airbnb and VRBO has dramatically expanded their availability and popularity. Private short-term accommodation activities are largely focused on the rental of houses, apartments, or rooms within residential properties. This growth in

E. Vaz, *Regional and Urban Change and Geographical Information Systems and Science*, Advances in Geographic Information Science,
https://doi.org/10.1007/978-3-031-24731-6_8

private short-term accommodation services has led numerous long-term rental properties to transition to short-term rentals. According to CBRE (2017), approximately 7-in-every-10 units listed on the Airbnb distribution platform are entire-home rentals.

Founded in 2008, the official Airbnb application was launched in 2010 which made transactions and booking more efficient; and by 2011, Airbnb expanded internationally and was offered in various cities across the globe (Deane 2021). Since its founding, Airbnb has had over 300 million guest check-ins worldwide (Airbnb, 2018a). Pre-COVID-19 Airbnb reported that they had 5.6 million global listings, with 150 million users worldwide (Airbnb 2018b). The United States is the largest Airbnb market, with $33.8 billion in collected revenue in 2020. France and Spain rank second and third on the list, respectively. Canada ranks at number six with a total collective revenue of $4.3 billion—an increase from $2.8 billion in 2018 (Airbnb Statistics 2020), accounting for roughly 18% of total traveler accommodation in Canada (CBRE 2017).

Over the past decade, short-term rental accommodations have become a "hot button" issue within many large cities in Canada and abroad. Indeed, cities like Toronto, Ontario (2020), and Vancouver, British Columbia (2020), have introduced bylaws to attempt to regulate the expansion of short-term rentals—in particular, Airbnb. Despite efforts by local governments to regulate the market, the pre-COVID-19 Airbnb listings continued to proliferate. As the housing advocacy coalition Fairbnb (2020, p. 2) argued:

> The results were shocking. We found that Airbnb would have had to remove 8,241 non-compliant listings. About 6,500 of these were found to be entire homes: houses, apartments and condos that had been permanently removed from the city's housing stock and used year-round to accommodate tourists and guests by so-called "high-volume" hosts.

Toronto is not unique, as similar analysis in Vancouver found that over half of listings contravene city regulations (Woodward 2020).

A key issue is that the rise of Airbnb has occurred at the same time as a rapid rise in housing and long-term rental costs. Toronto experienced a 145% rise in the monthly rental price of a single bedroom apartment between 2009 and 2019 (CMHC 2021). Similarly, the housing saw a rise in average sales price from $396,234 in 2009 to $819,153 in 2019 (TRREB 2021). This rise is in part due to the lack of vacancy. According to the Fairbnb (2021, p. 4), drawing on statistics collected by the CMHC (2019):

> There were 4,850 vacant primary units and 1,102 vacant condo rentals in Toronto in 2019. If even half of the approximately 7,300 non-compliant entire home listings were returned to the long-term rental market, the city's vacancy rate for condo and purpose-built rentals combined would ease, from 1.3% to 2%. Toronto's vacancy rate could approach a "healthy" rate of 3% if all non-compliant entire home listings became long-term rentals.

While Airbnb and other short-term accommodations are not the sole drivers of this market change, they have received considerable blame. As the Ontario Local Planning and Appeal Tribunal (LPAT 2019) argued: "One fact is indisputable: each dedicated short-term rental unit displaces one permanent household. That household must find another place to live." This issue is further heightened by multiple-property owners who have converted their properties from long-term residential units to short-term accommodations for tourists. The recent implementation of licensing bylaws in Toronto aims to reduce these impacts by prohibiting owners from listing properties as short-term rentals if it not their principal residence. In response to these changes, so-called 'ghost hotels'—where the landlord owns multiple units and doesn't actually live at the address listed on the site—will likely disappear from Toronto" (Thorben Wieditz, as quoted in Gallichan-Lowe 2020).

Within this context, this research aims to explore the changes in the Airbnb market in Toronto, Ontario, and Vancouver, British Columbia, to understand the way the market has changed in these two cities. In doing so, the study aims to answer two research questions:

RQ1: Are there spatial patterns in the distribution of Airbnb rentals in Toronto? And how have these changed between 2016 and 2019?

RQ2: Have these trends and patterns affected the long-term rental market in Toronto?

To answer these questions, two key methods are presented in this chapter: an emerging hotspot (EHA) analysis to track the changes in the distribution and concentrations of Airbnb and spatial regression to model the relationship between Airbnb presence and rental rates. In short, this research asks where Airbnb rentals are and what their impact is.

As noted, Toronto (population 2.93 million) will be the focus of this research. The city is the second most densely populated city in Canada with a population density of 4334.4 people per km^2. The total private dwelling was reported to be 1,179,100 (Statistics Canada 2016). The most common dwelling type was shown to be apartment buildings at 38%, followed by houses and condominiums at 28% and 20%, respectively (Statistics Canada 2016). Given the higher number of dwellings, Toronto has more potential to have a higher stock of short-term rentals. Recent statistics show that the average rent in Toronto is $1523 which is a 4.7% increase from the previous year (CMHC 2021). According to CBRE (2017), Toronto ranks as the second city with the highest number of Airbnb listings, while ranking first with the highest revenue compared to other Canadian cities. The data shows that over the span of 4 years (January 2016–December 2019), Toronto had 710,749 Airbnb listings (Inside Airbnb 2021).

8.2 Background

The rise in prominence of Airbnb is embedded within a larger context of technological innovation and sociological change, which has resulted in the proliferation of digital platform economies and sharing economies (Anwar 2018; Oskam and Bowsijk 2016). This is part of a larger transition for economies based on mass production, toward knowledge and value networks. Within these larger economic, technological, and societal shifts, Statistics Canada (2017) identified three prominent characteristics of these platform services. First, and most importantly, is the centralization of technology and how it is used to facilitate transactions and connect with a wider network of potential consumers. Second, the unique ability to rent and borrow products rather than buying and owning them; this has allowed individuals to leverage their existing assets to provide service in the market and diversify their sources of income. Lastly, the individualistic nature of market transactions between person to person is in high contrast to the traditional services provided by firms and companies to individuals (Statistics Canada 2017).

8.2.1 Digital Platform Economy and the Sharing Economy

The emergence of the sharing economy in the past decade has caused a significant shift on user consumption from traditional businesses and services to digital business models. Similar terms such as platform economy, collaborative economy, or on-demand services have been used interchangeably to describe this emerging area of economic activity. Through the sharing economy, services are often provided through an online platform, where the consumers can purchase goods and services digitally.

From a technological perspective, new communicative technologies facilitate greater connectivity through peer-to-peer networks. Within this new configuration, Rifkin (2014) uses the term "sharing economy" to describe a hyperconnected economy. This differs from traditional conceptions of economy—where marketplaces connect supply and demand between customers and companies—as this new configuration leverages digital platforms connect individual consumers to each other (Sovani and Jayawardena 2017). Frenken et al. (2015) add further perspective, arguing that the sharing economy as consumers grants each other temporary access to underutilized physical assets, such as residences (or in the case of companies such as Uber, vehicles). This "idle capacity" (Oskam and Bowsijk 2016, p. 25) can now be leveraged more freely, often for money from other consumers.

A second factor driving the rise of the sharing economy is the changing concept of value within society (Brand and Rocchi 2011; Guttentag et al. 2018). This change is broadly described as a transition from an industrial, material-based economy that emphasizes product ownership to one that is much more experiential. As Oskam and Bowsijk (2017, p. 23) note, this change presents as shift to a "knowledge economy with focus on self-actualization towards a transformational economy with focus on a higher purpose and searching for meaningful living."

8.2.2 Housing and Rental Market in Canada

One of the main areas of research on the topic of home-sharing and short-term accommodations is its relevance to the housing and rental market. A key reason for this impact is that Airbnb is viewed by consumers as an alternative to traditional hotel accommodations. Zervas and colleagues (2016) suggest that the entry of Airbnb into the accommodation sector has impacted the hotel revenue by 8–10%. The authors attribute this to periods of high demand and the platform's ability to provide instantaneous supply to scale to meet demand. Similarly, Guttentag et al. (2017) show that two-thirds of Airbnb guests use the service as a viable substitute to traditional accommodations.

The main concern highlighted by various researchers and the public is that outcome of this adoption of Airbnb and their proliferation in the market decreases the number of long-term rental units. While this may not be a concern to home buyers, it poses a challenge to the renter population, where short-term rentals such as Airbnb are inevitably affecting the supply of long-term rental units. Housing and rental affordability has been a rising issue in Canada even prior to the emergence of digital platform economy and Airbnb.

Affordability measures the proportion of income renter households spent on rent and utilities, and housing is considered affordable if the household spends less than 30% of its before-tax income on rent and utilities. Over the past decade, Vancouver, Toronto, and Montreal have been coined as the cities with the highest housing and rental prices, and property values. This trend has been consistent over the recent years; the 2016 Canadian census shows that the average monthly rent in Toronto was listed as \$1242. By 2020, Toronto reported an average monthly rent of \$1523 which represented a 4.7% year-over-year increase (CMHC 2021). This rise in cost is mirrored by high demand—only about 3% of rental units in Toronto were vacant in 2020 (CMHC 2021). In Toronto, 46% of households spent more than 30% of their income on rent and utilities, while 21% of households spent more than 50% of their income on rent. Though this chapter focuses primarily on Toronto, similar statistics were found in Vancouver, where 43%, and 21% of households spent more than 30% and 50% of their income on rent and utilities, respectively (Canadian Rental Housing Index 2016). These numbers show that over half of the household populations in both cities are facing housing affordability issues.

The association between short-term rentals such as Airbnb and rent prices has been studies by numerous researchers. The key findings suggest that the rapid growth of short-term rentals has increased issues around availability and affordability of housing in many global cities. In relevance to the Canadian housing and rental market, scholars suggest that the growth of Airbnb is likely converting units of housing from the long-term rental market to the short-term rental market. This pattern is evident in Montreal, Toronto, and Vancouver and is in large due to the increase in the commercialization of Airbnb operators (Wachsmuth et al. 2017). Moreover, studies suggest that the distribution of Airbnb has shown to negatively impact the housing market both in Montreal and Toronto (Desmirarais 2016; Wieditz 2017).

The increase in STRs over the years has facilitated the conversion of housing supply from long-term to short-term rentals while increasing the economic value of properties dedicated to STRs (Combs et al. 2019). Barron et al. (2019) analyze the overall effect of home-sharing on housing prices and rents in the United States and found that 1% increase in Airbnb listings leads to a 0.018% increase in rents mainly due to the reallocation of the housing supply from the long-term rentals to short-term rental units. Similarly, the doubling of Airbnb activities has shown to increase property values between 6% and 11% in New York City (Sheppard and Udell 2018). The presence of Airbnb and its contribution to increased rental rates appear to be prominent in numerous international cities, such as London (Shabrina et al. 2021), Ireland (Lima 2019), Barcelona (Garcia-Lopez et al. 2019), Los Angeles (Lee 2016), and Boston (Horn and Merante 2017).

Aside from the presence of Airbnb and its contributions to high rental rates, researchers suggest that the rapid misuse of Airbnb units is an additional contributor to increased rent prices. This occurs when hosts list more than one offer, which has been shown in cities like London and Berlin, where more than 2% and 0.30% of all properties, respectively, were misused through Airbnb as short-term rentals (Shabrina et al. 2021; Schafer and Braun 2016). Moreover, the misuse of housing properties as Airbnb holiday rentals was shown to be geographically concentrated in areas with high private rental stock (Shabrina et al. 2021). This further emphasizes how multiple-property owners are motivated by the financial incentive of listing owned properties as short-term holiday rentals. In 2016, the ownership rate in the City of Toronto was reported to be 66.5% with a renter share of 33.5% which is lower than the national rate of 67.8% (Statistics Canada; Census 2016). This shows that there is a high percentage of house owners who could potentially be participating in Airbnb activities.

More recently, data released by the Canadian Housing Statistics Program (2018) shows nearly half of the multiple-property owners who lived in the City of Toronto also owned other properties within the same geographic area/census subdivision. In Toronto, the share of multiple-property owners of all owners was reported to be 16.4%. This implies that almost one in every five property owners in both cities owns more than one property (Bekkering et al. 2019). This factor, along with high population density and dwelling type density, contributes to the high number of Airbnb listings in the City of Toronto.

Other relevant studies point to the complex social issues surrounding the expansion of peer-to-peer rentals such as Airbnb. Past research has highlighted the connection between short-term rentals and gentrification (Mermet 2017; Wachsmuth and Weisler 2018). This concept has been described as the "short-term rent gap" by Wachsmuth and Weisler (2018), whereby the financial incentive of short-term rentals drives the displacement of low-income residents in favor of tourists. Similarly, Yrigoy (2019) notes that the conversion of residential rentals to tourist rentals introduces a "potential ground rent," whereby as the number of Airbnb listings increases, housing affordability for residents decreases, and therefore increasing the risk of displacement. Other studies have examined the sociodemographic and transportation factors may affect Airbnb listings, noting that Airbnb rentals are more likely to be located in neighborhoods closer to the city center while having good access to transit services, high median house value, and a high median household income (Jiao and Bai 2020). Collectively, these studies suggest that the presence of Airbnb may inevitably contribute to issues surrounding social inequality while negatively affecting vulnerable population groups.

Finally, these proliferations of Airbnb and other short-term rentals do not appear to be homogenous across urban spaces, but rather are clustered. Indeed, several studies have also examined the price and location of Airbnb listings in relation to proximity to areas with tourist amenities (Nelson 2010; Tsai et al. 2016). Findings suggest that the presence of amenities has a positive impact on the sale value of Airbnb listings. Gutiérrez et al. (2017) also found similar patterns of distribution in Barcelona, where Airbnb was shown to benefit more from being near sightseeing spots when compared to the traditional hotel accommodations. The authors also note that Barcelona's Airbnb listings are concentrated in the city center which further contributes to the tourism pressure over residential areas (Gutiérrez et al. 2017). More relevant to the Canadian context, Gibbs et al. (2017) examine how Airbnb sets nightly prices. Findings show that price of listings is significantly impacted by physical characteristics, location, and host characteristics.

8.2.3 Policy and Regulation

In contrast to other areas of research in the context of short-term rentals, very few studies have examined the regulation and policies surrounding peer-to-peer rentals. Several reports have addressed the regulation of short-term rental in the Canadian context while drawing from empirical analysis in other countries (Jamasi 2017; Wachsmuth et al. 2017). However, due to the lack of regulations and policies, there are no studies that have explicitly examined the Canadian regulation and policies around the short-term rental markets.

Researchers have suggested policy recommendation based on empirical analysis of Airbnb regulation in other cities. The most notable is the work by Wachsmuth et al. (2017) in which they highlight three recommendations based on the performance of regulations implemented in Amsterdam. In 2017, the City of Amsterdam and Airbnb made an agreement whereby the platform automatically blocks booking of entire-home listings once 60 days of bookings has been reached. This measure, alongside strictly permitting listings that are one's primary residence, has significantly reduced the commercialization of Airbnb and disincentivizes multiple-property owners from listings multiple units. Based on these results, Wachsmuth et al. (2017) highlight three recommendations for cities in Canada. First, a single host is only prohibited to list a single rental. Second, the number of days an entire home can be rented is limited to reduce the full-time short rentals. Third, the establishment of an agreement with the platform would allow for a more efficient tracking of regulations and the compliance of Airbnb hosts.

In the Canadian context, the City of Vancouver, British Columbia, was the first to make some progress in regulating short-term rentals by implementing the license bylaw, requiring that all short-term rental operators must have a business license and include their license number in all online listings (City of Vancouver 2020). Following suit, Toronto has also taken steps toward regulating STRs by requiring all operators to register with the city and list their registration number in all online listings. Moreover, all operators are required to collect and remit a 4% Municipal Accommodation Tax (MAT) on all listed rental units (City of Toronto 2020).

8.3 Methodology

The objectives of the research presented in this chapter were to understand how the spatial distribution of Airbnb listings changed between 2016 and 2019 in Toronto, Ontario, and to understand the effect that it had on the long-term rental rates in the city. To account for the COVID-19 pandemic, data from January 2020 onward was omitted as it had significant impact on the Airbnb market in Canada. To explore these issues, a multi-method approach was used: first, an emerging hotspot analysis was used to identify spatial-temporal patterns of Airbnb listings; and second, spatial regression modeling was used to explore the relationship between concentrations of Airbnb listings and rental rates.

This approach has several benefits. First, it fills in existing research gaps as previous research into short-term rentals has focused on the growth of the Airbnb market (Combs et al. 2020), explored local attributes that impact short-term accommodation prices (Deboosere et al. 2019), its impacts on neighborhood (see Wachsmuth and Weisler 2018), and its impact on housing and rental markets (Grisdale 2018; Wachsmuth et al. 2017). This research extends existing scholarship as it adds a temporal element to the analysis. Second, the use of spatial analysis provides robust methods for exploration and insight into geographic processes (Burt and Barber 1995; Rogerson 2010) and allows for hypotheses to be constructed prior to data collection and for these hypotheses to be tested and validated (Burt and Barber 1996). This method also allows for the identification of cause-and-effect relationships and the ability to generalize research findings drawn from a large population (i.e., all census tracts of Toronto). In addition, the research design may be able to eliminate the confounding of spatial autocorrelation between variables (Rogerson 2010).

8.3.1 Data

The research presented in this chapter relies on three key pieces of data (summarized in Table 8.1): Airbnb location data, boundary data, and census data. The Airbnb location data was acquired from Inside Airbnb.[1] Data was acquired based on monthly listings from January 2016 to December 2019. Relevant attributes such as latitude and longitude, price, and property type were extracted and sorted into a tabular format, which was then geocoded in ArcGIS Pro. It is important to note a

[1]Airbnb listing data was acquired though Inside Airbnb which is an independent and noncommercial set of tools that allow users to explore the distribution of Airbnb in cities around the world.

Table 8.1 List of variables used in the analysis to address research questions

Research question association	Variables	Data sources
Q #1	Airbnb listings (2016–2019)	Inside Airbnb, 2020
	Number of dwellings (2018–2019)	Environics, 2020
Q #2	Airbnb listings (2016–2019)	Inside Airbnb, 2020
	Average expenditure on rent (2018, 2019)	Environics, 2020
	Number of owners (2018, 2019)	Environics, 2020
	Number of renters (2018, 2019)	Environics, 2020
	Number of dwellings (2018–2019)	Environics, 2020

limitation of the Airbnb data, which is that location information for listings is anonymized by Airbnb. In essence, this means that location for a listing on the map is offset by 150 m from the actual address.

As noted, the Airbnb location data was aggregated for this analysis. Census tracts—delineated by Statistics Canada for the 2016 census—will be used as the spatial unit of analysis. Analysis at the census tract level provides more granular information on the potential impact of Airbnb listings on average rent values. Toronto is divided into 572 census tracts.

The dependent variable in this study was average total rent expenditure which was directly acquired through SimplyAnalytics for the year 2018 and 2019. Additionally, ownership ratio was included in the analysis as a control variable and was calculated by dividing number of owned dwellings by number of rented dwellings.

8.3.2 Analytical Approach

8.3.2.1 Emerging Hotspot Analysis

The first research question is centered around identifying the emerging hotspots in Toronto. This process was completed for each city individually by using emerging hotspot analysis (EHA), whereby the spatial pattern of Airbnb was analyzed over space and time to determine the change in significance over the four-year study period (January 2016–December 2019).

EHA is classified as a type of space-time pattern mining, where trends are identified in the clustering of point densities in a space-time cube created by either aggregating points or setting defined locations. In this case, Airbnb locations were aggregated at the census tract level (space) to identify areas with consistent pattern over time (i.e., by month). The spatial analysis component uses the Getis-Ord Gi* statistic to determine statistical significance. Features are analyzed alongside neighboring features; however, a feature with a high value is not automatically classified as statistically significant. To be a statistically significant hotspot, a feature will have a high value and be surrounded with other features with high values as well. The local sum for a feature and its neighbors is added up and compared to the proportion of sum of all the other features. Therefore, if the local area and the sum of the values in that local area are very different from the average and proportional value of all the data across all the areas, and the difference is calculated to be a significant rather than a random result, then the area is classified as either a significant hot or coldspot. Gi* values were calculated for each census tract for each month of the study period (so 48 values or "time slices" were calculated for each of the 572 unique areas in Toronto). The temporal analysis uses the Mann-Kendall trend test which assesses whether the Gi* for each census tract was increasing over time or decreasing over time, and whether the trend in either direction is statistically significant. Using these two analysis tool—one to determine spatial autocorrelation and one to determine how it behaves over times—the EHA categorizes each spatial unit of analysis (i.e., census tracts) into one of 17 categories. These include new, consecutive, intensifying, persistent, diminishing, sporadic, oscillating, and historical hot and coldspots, or can be shown as no pattern detected. The EHA was conducted in two ways: first, it was done for the entire study period (January 2016 to December 2019); and second, it was conducted for each year of the study. For the entire study period, a z-score was calculated for the averages of 2019 rental expenditures and percentage of renters for each category of hotspot that emerged, to determine if hotspots were associated with higher rental rates and expenditure costs.

8.3.2.2 Spatial Regression

In the second part of the study, spatial regression was used to understand the influence of Airbnb concentrations on rental rate. To do so, rental expenditures (dependent variable) were modeled against Airbnb rental rates (independent variable) at the CT level in Toronto. Based on existing research on short-term accommodations (see Grisdale 2018; Wachsmuth et al. 2017) and the outcome of the emerging hotspot analysis, it was strongly suggested that there may be spatial relationships within the locations of Airbnb.

Nonspatial regression analysis typically utilizes an ordinary least squares (OLS) estimator to model the relationship between dependent and independent variables. The regression model can be defined as:

$$y = x\beta + \varepsilon \tag{8.1}$$

where y is the response variable, x is the explanatory variable, and β is the estimate of the coefficient. The error term (ε) is assumed to be normally distributed ($\varepsilon \sim N(0, \theta^2)$).

However, a key assumption of OLS is that there is no autocorrelation between observations of the variable. This estimator implicitly assumes that each region is independent of all other regions (Baumont et al. 2001). However, if there is spatial dependence, the endogeneity of the regressors causes the OLS estimator to become biased and inconsistent and produces inefficient estimators and the potential for invalid statistical inference (Abreu et al. 2005).

To ensure that an appropriate methodology was used, an analytical approach laid out by Anslin (1994, 2001) and Cleave et al. (2020) was utilized. In it, a weights matrix (based on Queen's contiguity) is included in the OLS regression model to test for spatial autocorrelation. To do so, Moran's I and Lagrange multiplier (LM) were used as diagnostics. If Moran's I was not significant, it indicates there is no spatial autocorrelation present in the dataset and a standard OLS regression model is adequate. If autocorrelation is present, then the LM diagnostics were used to test for spatial lag and error. Spatial lag is where dependent variables are influenced by observed independent variable values in neighboring areas. Alternatively, spatial error is where there is correlation in the error between neighboring regions. For the first two instances when OLS assumptions are broken, a maximum likelihood (ML) estimator is needed to produce a valid model between dependent and independent variables. To show these patterns within the data, ML error and spatial lag were modeled. The model for spatial lag is:

$$y = \rho Wy + \beta x + \mu, \qquad \mu \sim N\left(0, \theta^2\right) \tag{8.2}$$

where ρ is the lag coefficient and Wy is the weighted matrix. The model for spatial error is:

$$y = \lambda W\varepsilon + \beta x + \mu, \qquad \mu \sim N\left(0, \theta^2\right) \tag{8.3}$$

where λ is the error coefficient and Wε is the weighted matrix. The model was selected based on whether the LM diagnostics indicated only one of lag or error present. If the initial LM tests indicate the presence of both, then robust LM tests were used to further evaluate the model suitability.

This process ultimately creates 12 models. First, the analytical approach was used to compare the average rental expenditure in each CT with the Airbnb rate (average number of monthly listings divided by total dwellings in the CT) in a bivariate regression. This was followed by a multivariate regression that included Airbnb rate and the ratio of homeowners to renters (as a control variable). This was done for 2018 and 2019 for both study areas. This was followed by models that compared the rate of change in rental expenditure between 2018 and 2019 and the change of the Airbnb rate in the same period. This analysis was limited to these years, as rental expenditure data was only available for 2018 and 2019.

8.4 Results

Table 8.2 provides an overview of the characteristics of Toronto during the study time period. The rapid growth of Airbnb is evident in both cities from 2016 to 2019. In Toronto, the number of Airbnb listings rose by 115% from 2016 to 2019 (108,400 to 232,600 listings).

Table 8.2 Descriptive summary for City of Toronto

City	Total number of Airbnb listings (1000)				Renter (%)	Ownership (%)	Annual rent expenditure in 2019 ($)
	2016	2017	2018	2019			
Toronto	108.4	159.8	186.3	232.6	44	56	$18,276

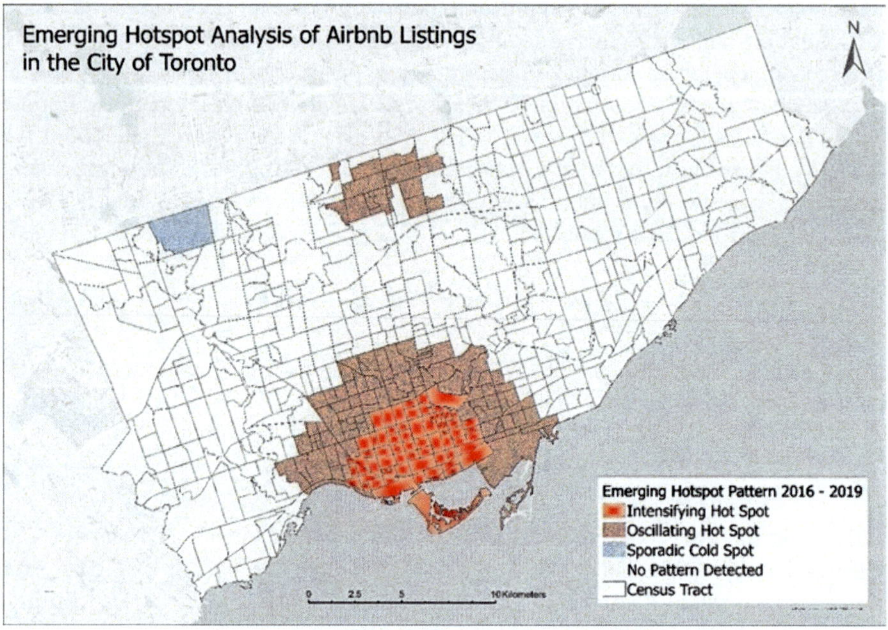

Fig. 8.1 Emerging hotspots for Airbnb listings in the City of Toronto at the census tract level from January 2016 to December 2019. (Data source: Inside Airbnb, 2021)

8.4.1 *Emerging Hotspot Analysis*

Figure 8.1 presents the Airbnb hotspots at the census tract level that occurred between 2016 and 2019 based on the EHA. The clustering of the intensifying hotspots in the downtown core shows that majority of listings have been geographically concentrated in these census tracts. Intensifying hotspots refer to locations that have been statistically significant hotspots for 90% of the time-step intervals (monthly-basis). Moreover, the intensity of clustering of high counts of listings during each month is increasing overall and that increase is statistically significant. Surrounding these intensifying hotspots are census tracts that are classified as oscillating hotspots which indicates that these areas have been statistically significant hotspots for the final time-step interval but also have been statistically significant coldspots in prior time steps. In contrast, one census tract in the peripheral edges of the city has been identified as a sporadic coldspot. This implies that this census tract oscillates on and off as a coldspot with no prior history of being a significantly hotspot.

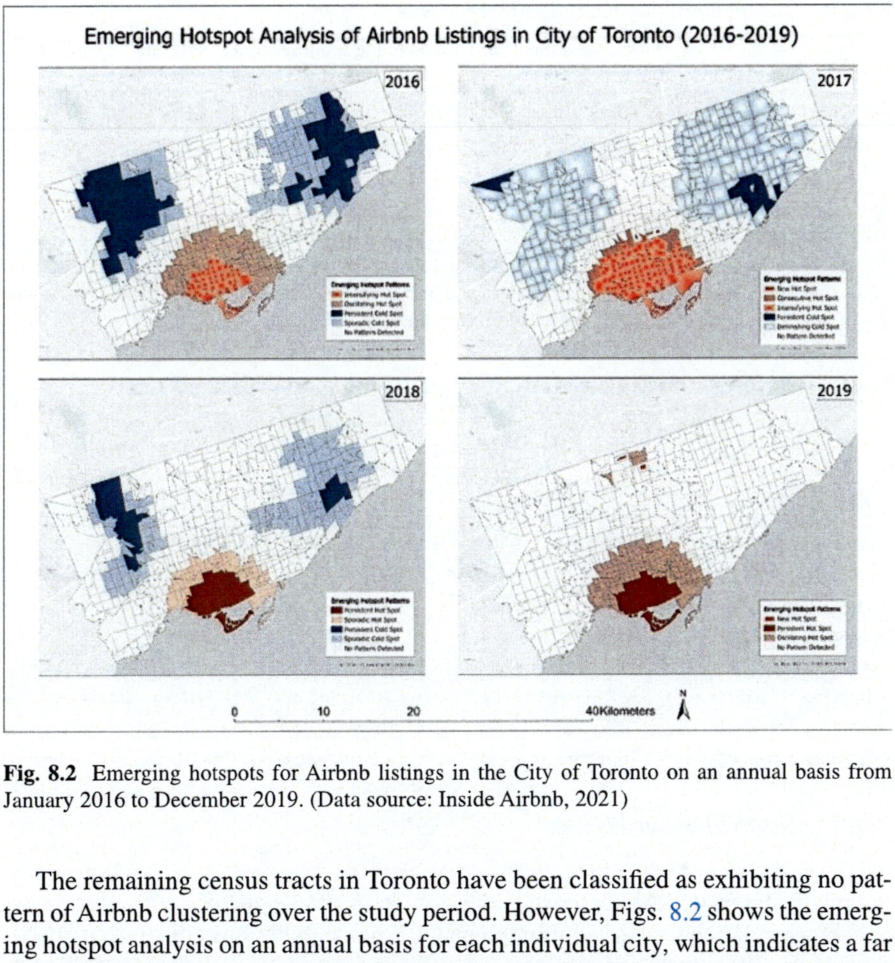

Fig. 8.2 Emerging hotspots for Airbnb listings in the City of Toronto on an annual basis from January 2016 to December 2019. (Data source: Inside Airbnb, 2021)

The remaining census tracts in Toronto have been classified as exhibiting no pattern of Airbnb clustering over the study period. However, Figs. 8.2 shows the emerging hotspot analysis on an annual basis for each individual city, which indicates a far more variable hotspot patterns in different areas of the city. Of note, there is an intensifying hotspot in the core of the city in 2016 and 2017, which becomes a persistent hotspot in 2018 and 2019. This suggests an expansion occurred in the first half of the study period, though stabilizing in the second half still represented the area with the highest concentration of listings in Toronto.

Table 8.3 summarizes the 2019 housing characteristics for the emerging hotspot classes in Toronto. The 70 census tracts that were identified as emerging hotspots were found to have significantly higher rates of renters (63%) and higher annual costs than the city average ($21,991 per year). The 100 oscillating hotspots were also found to have statistically significantly higher rental costs ($19,767). For Toronto, it is evident that the percentage of renters are concentrated in the downtown center where majority of the Airbnb listings are clustered as well. Census tracts closer to the downtown core appear to have a higher percentage of home renters and are Airbnb hotspots. Conversely, census tracts in the peripheral of the city boundary tend to have a higher percentage of homeowners and are Airbnb coldspots.

Table 8.3 Description of hotspot classes in the City of Toronto

Hotspot class	Average annual rent ($)	Average monthly rent ($)	Renter %	Owner %
Intensifying hotspot (70)	21,991*	1833	63*	37
Oscillating hotspot (100)	19,767*	1647	49	51
Sporadic coldspot (1)	16,825*	1402	34*	66
No pattern (401)	17,255	1438	42	58

Notes: * significant difference from city average at $\alpha = 0.05$ based on z-score (+/– 1.96). Mean for annual rent expenditure is $18,276 (st. dev of $627); average renter % is 46% (st. dev of 3%)

Table 8.4 Regression diagnostics

	Bivariate			Multivariate		
	2018 Toronto	2019 Toronto	Toronto change	2018 Toronto	2019 Toronto	Toronto change
Moran's I	0.386***	0.397***	0.012	0.386***	0.397***	0.012
LM (lag)	242.69***	260.41***	44.49*	248.76***	276.34***	0.50
LM (error)	245.75***	271.97***	37.34*	277.35***	342.13***	0.39
R-LM (lag)	1.56	0.29	**9.36****	1.06	1.37	–
R-LM (error)	**4.62***	**11.85****	9.67*	**29.65*****	**67.15*****	–
Model used	ML (error)	ML (error)	ML (lag)	ML (error)	ML (error)	OLS

*significant at $p < 0.05$; **significant at $p < 0.001$; ***significant at $p < 0.0001$

8.4.2 *Spatial Regression*

Following the analytical approach laid out in the methods, the presence of the spatial autocorrelation was tested through an OLS model with a weights matrix incorporated (results summarized in Table 8.4). Based on the analysis, one model did not show spatial autocorrelation (rent change in Toronto—multivariate). This is a notable finding, as it suggests that rent change does not have the same underlying structure across the study areas as average rent does. Only one model—the bivariate rent change in Toronto model—was found to have spatial lag, indicating there was some spatial dependency in rent change across the city. The other models (2018 and 2019 rent—both bivariate and multivariate) were found to have spatial autocorrelation in the error. This is unsurprising since only two variables were included in the regression, meaning there is potential for other variables—which may be correlated—that may also help explain variation in rental rates.

Table 8.5 summarizes the bivariate models for Toronto. It was found that there was a significant positive correlation between Airbnb presence and the rent level. Additionally, there was a significant relationship between rent change and change in Airbnb prevalence. In basic terms, this means that the bivariate analysis showed that greater Airbnb in a census tract is associated with higher levels of rent expenditure.

Table 8.5 Results of the bivariate analysis, examining the relationship between Airbnb presence in 2018 and 2019 and average rent expenditure in the City of Toronto

Year	Independent variables	Coefficient ($ in rent)	Std. error	R^2
2018	Airbnb rate (rentals/dwellings)	815.29**	270.64	0.376
2019	Airbnb rate (rentals/dwellings)	679.67*	325.41	0.334
Change	Airbnb rate change (2018–2019)	756.43*	390.44	0.07

*significant at $p < 0.05$; **significant at $p < 0.001$

Table 8.6 Results for the spatial regression examining the relationship between Airbnb listings and average rent expenditure in the City of Toronto (N = 572)

Year	Independent variables (Toronto)	Coefficient ($ in rent)	Std. error	R^2
2018	Airbnb rate (rentals/dwellings)	577.94**	304.76	0.383
	Owner to renter ratio	1004.66*	105.18	
2019	Airbnb rate (rentals/dwellings)	667.84*	99.60	0.497
	Owner to renter ratio	1139.21**	245.56	

*significant at $p < 0.05$; **significant at $p < 0.001$

Table 8.6 summarizes the findings of the multivariate analysis. In 2018, it was found that both Airbnb rate ($1004.66; $p < 0.001$) and the owner-renter ratio ($577.94; $p < 0.05$) had statistically significant positive relationships with rent. This means that as the number of houses owned increased (and by extension the number of rental properties decreased) and the number of Airbnbs increased, so did rent. This suggests that even in CTs with a high demand for long-term rentals, Airbnb's presence is influencing rent. A similar pattern was observed in 2019, where both Airbnb rate ($667.84; $p < 0.001$) and owner-renter ratio ($1139.21; $p < 0.05$) again were found have a statistically significant relationship with rent. The model produced an R^2 of 0.383 and 0.497 for the 2018 and 2019 base year, respectively. This implies that in both years, approximately between one-third and one-half of the variation in average expenditure on rent can be attributed to the independent variables included in the model.

Finally, Table 8.7 summarizes the results of the model comparing change between 2018 and 2019. For Toronto, the intercept was $1002.95, indicating a rent increase occurred if no other change to Airbnb rate or ownership changed. In Toronto, it is noted that a positive change in Airbnb rate had a significant positive relationship with rent. However, ownership was not significant. Still, this suggests that increased presence of Airbnb is linked with increased rental rates.

In this model, there are several other key things to note. First, spatial autocorrelation was not detected in the diagnostics for either city, so an OLS estimator was used (differing from the yearly models that used ML estimators for error). This suggests that the change in rent was much more "random"—potentially due to the short timeframe of the analysis, which prevented trends like turnover or gentrification or demand to really influence the change. Additionally, the model explained far less variation (i.e., R^2; see Table 8.7) compared to the yearly models. This suggests that at least in this time period, there were other factors that influence rent change that were unaccounted for.

Table 8.7 Results for the spatial regression examining the relationship between Airbnb listings change and average rent expenditure change

City	Independent variables	Coefficient ($ in rent)	Std. error	R^2
Toronto	Airbnb rate change	736.00	392.11*	0.103
	Owner to renter change	25.21	197.99	

*significant at $p < 0.05$; **significant at $p < 0.001$

There are several key trends across this analysis that are important for understanding the relationship between Airbnb presence and rent. First, in all models, the relationship between Airbnb rate change and rental rate was significant and positive. This indicates that the presence of Airbnbs is related to higher rents. It can be interpolated that Airbnbs are not just locating in areas of high rent (which itself, is likely at least a partial explanation) but that there is evidence that increased short-term rental intensification was affecting rent.

Additionally, the annual models show a positive relationship between owner to renter ratio and average expenditure on rent. This makes sense, as areas with higher ownership rates are likely to have higher demand for the rental properties, which could increase their price. However, if framed as a control variable—where owner to renter ratio acts as a proxy for demand—it provides further evidence for Airbnb positively affecting rent, as that relationship also remained positive and statistically significant.

8.5 Discussion

The research presented in this chapter considered two key questions. The first research question in this study focused on identifying changing patterns of Airbnb listing in the City of Toronto from 2016 to 2019 using an emerging hotspot analysis. As illustrated by the result of the study, the majority of Airbnb listings are evidently concentrated in the downtown area in Toronto. Typically, the downtown core comprises a high population density and a high dwelling density. Over the years, this trend has been persistent in both cities, where the population growth, and therefore housing growth, has been geographically concentrated in the inner-city core (Jiao and Bai 2020). Smaller areas closer to the city center area are often characterized by a housing stock which consists primarily of multiple dwellings, condominium, and apartment buildings. The common dwelling type in these areas is one of the key indicators of the high distribution of Airbnb listings over the four-year study period. This is consistent with existing literature, where the ideal dwelling type for Airbnb listings has shown to be apartment buildings compared to houses (Krause and Aschwanden 2020). This is in large attributed to the competitive revenue and the number of Airbnb reservations. Apartment buildings are typically shown to generate more revenue when listed as a short-term holiday rental than they would when listed as a long-term rental unit. Moreover, apartments have shown to have more flexibility around the number of guests and minimum stay which increases the booking preference of apartments as short-term holiday accommodations (Krause and Aschwanden 2020).

The tourism characteristics of these areas are another key driver of the high Airbnb listing concentration. The city center in Toronto is comprised of a variety of tourist attractions including museums, aquarium, botanical garden and parks, beaches and bodies of water, and lookout spots such as the CN Tower (Gutiérrez et al. 2017). Proximity to other amenities such as transportation services, diverse cuisines, and restaurants is also a key determinant of Airbnb location and guest attraction (Nelson 2010; Tsai et al. 2016). The downtown core in both Toronto and Vancouver has a high access to a network of transportation services such as subway and bus stations.

On a different note, the professionalization of Airbnb is one of the key drivers of the expansion and the rapid growth of the peer-to-peer rentals. The financial incentive through revenue generated by properties listed as short-term holiday rentals has facilitated the decline in the availability of long-term rental units. In other words, multiple-property owners gain a higher revenue when listing their properties in the short-term rental market. This is especially evident in Toronto, where homeownership and multiple-property ownership are higher relative to other Canadian cities (CHSP 2018; Bekkering et al. 2019). Moreover, the city center has shown to be home to highly skilled individuals with high pay (Heisz and Larochelle-Côté 2005). This sheds some insight on the socio-economic characteristics of population groups with ownership of multiple properties closer to the downtown core.

The second research question focused on the link between Airbnb presence and rent expenditure levels. The regression analysis suggests that the average number of Airbnb listings has a positive relationship with average expenditure on rent. This was evident on an annual basis for 2018 and 2019 in Toronto. The results show that the average number of Airbnb listings is positively correlated with the average expenditure on rent. These findings are consistent with previous studies such as those by Lima (2019), Lee (2016), and Horn and Merante (2017), where the presence of Airbnb has shown to contribute to the increase in rental rates. This relates back to the issues surrounding the availability and affordability of housing in the long-term rental market. The rapid conversion of rental units to short-term holiday rentals increases the pressure around the shortage of available housing. It further highlights how the growth of Airbnb has contributed to a decrease in housing supply by reallocating residential units from the long-term rental market to the short-term rental market (Barron et al. 2019).

The findings of this analysis align with existing research, where scholars have shown that there is a positive correlation between the presence of Airbnb and rental rates (Sheppard and Udell 2018; Shabrina et al. 2021; Comb et al. 2019). The patterns in Toronto appear to be similar to those observed in London (Shabrina et al.

2021), Ireland (Lima 2019), Barcelona (Garcia-Lopez et al. 2019), Los Angeles (lee 2016), and Boston (Horn and Merante 2017). The results of both the bivariate and the multivariate analysis show that a positive change in Airbnb rate had a significant positive relationship with rental rates in Toronto. These findings are similar to that of studies by Combs et al. (2019) where Airbnb was shown to be unevenly concentrated in urban centers and large cities such as Toronto, Montreal, and Vancouver.

Airbnb is only an additional variable to the already existing housing and rental crisis in Canada. Data shows that housing and rental prices are on a continuous rise across Canadian cities. This trend is more so evident in large cities such as Toronto, Vancouver, and Montreal, where the shortage of affordable housing supply continues to increase (Desmirarais 2016; Wieditz 2017). Therefore, it is even more critical to prioritize policies that bring forth the regulation of short-term rentals. The recent implementation of license bylaws in Toronto is the first step in regulating and tracking short-term rental activities across the city. This would allow policy makers to identify areas where Airbnb activities may be disrupting the local housing and rental markets. These regulations also help to reduce the misuse of Airbnb units which has shown to further contribute to increased rent prices (Shabrina et al. 2021; Schafer and Braun 2016). It is imperative that policies focus on identifying the distinction between short-term operators who are part-time home sharers and commercial operators (Combs et al. 2019). Doing so would allow industry experts to implement policies that slow the conversion of long-term rentals to short-term accommodations while maintaining the stock of housing market.

This study acknowledges that high rental rates and housing costs are products of complex and multifaceted issues. The linear regression analysis shows how the spatial distribution of Airbnb could impact the rental expenditure over time. Although the impact has shown to be relatively small, it further contributes to the larger complexities and issues around housing availability and affordability. The connection between Airbnb listings and average expenditure on rent could be used to identify vulnerable population groups prone to gentrification. This aligns with existing research which highlights the connection between short-term rentals and gentrification, whereby the conversion of residential rentals to tourist rentals increases the displacement of low-income residents while decreasing the stock of long-term rentals (Merme 2017; Wachsmuth and Weisler 2018). This cycle could disproportionally impact marginalized groups including recent immigrants, racialized communities, and low-income households (Heisz and Larochelle-Côté 2005). This is especially important in the case of Toronto, which is home to high percentage of immigrants. Future research is needed to gain a better understanding of the characteristic of population groups that are negatively affected by the growth of the short-term rental market.

8.6 Conclusion

This chapter provided a valuable and nuanced analysis of how Airbnb listings are spatially distributed in Canada's largest city over the period of 4 years. Using novel methods such as the emerging hotspot analysis, this study highlights how Airbnb listings are geographically concentrated in the inner-city core which is primarily motivated by dwelling type and revenue. It further explores how the average number of Airbnb listings is positively correlated with rent expenditure. Overall, it contributes to the growing body of research on the domain of digital platform economy and the short-term rental accommodations. It sheds light on the nature of these services within the Canadian context which could be insightful to researchers and policy makers for future scholarly and policy research.

Recall that the literature on the topic of Airbnb and its impact on the housing and rental sector as well as the tourism sector is relatively scarce. While this study focused on the direct relationship between Airbnb listings and rent prices, it would be insightful to examine how other externalities such as the COVID-19 pandemic have affected this relationship. As evident in recent reports, the closure of international borders has significantly decreased tourist activity and migration to large cities and therefore increasing vacancy rates while lowering hotel and Airbnb activities. A pre-pandemic and post-pandemic comparison would shed light on the resiliency of the peer-to-peer market as well as the housing and rental market and how they have been impacted by these external volatilities.

Several studies highlight the geographically driven aspect of Airbnb concentration in large cities. Potential studies could examine the difference between rural areas and sub-urban and urban areas. Drawing from existing research, it would be interesting to examine the concentration of Airbnb listings in different international cities with high tourism activity. This type of analysis could be further used to examine the flow of tourists or newcomer populations within a city. Moreover, further research on the sociodemographic characteristics of Airbnb hosts and Airbnb guests would allow for a better understanding how different population groups are affected through the spatial distribution of Airbnb. This applies to the tourism sector, as well as the existing tenants in the long-term rental market. These population groups may be especially prone to external impacts such as loss of employment or displacement through the expansion of Airbnb. This would further contribute to the research on policies and regulations regarding peer-to-peer accommodations and may facilitate the implementation of policies to regulate the activity of the short-term rental accommodations while protecting vulnerable populations.

References

Airbnb (2018a) Airbnb unveils roadmap to bring magical travel to everyone [Press release]. Retrieved fromhttps://press.atAirbnb.com/Airbnb-unveils-roadmap-to-bring-magical-travel-to-everyone/

Airbnb (2018b) Fast facts. Retrieved from https://press.Airbnb.com/fast-facts/

Airbnb Statistics (2020) iProperty management: data and research. https://ipropertymanagement.com/research/airbnb-statistics

Anwar ST (2018) Growing global in the sharing economy: lessons from Uber and Airbnb. Glob Bus Organ Excell 37:59–68

Barron K, Kung E, Proserpio D (2019) Research: when Airbnb listings in a city increase, so do rent prices. Harv Bus Rev 17:1–62. https://hbr.org/2019/04/research-when-airbnb-listings-in-a-city-increase-so-do-rent-prices#

Bekkering E, Deschamps-Laporte JP, Smailes M (2019) Housing statistics in Canada, residential property ownership: Real estate holdings by multiple-property owners. In: Statistics Canada. Retrieved from https://www150.statcan.gc.ca/n1/pub/46-28-0001/2019001/article/00001-eng.htm

Brand R, Rocchi S (2011) Rethinking value in a changing. A model for strategic reflection and business transformation, Philips Design, Amsterdam

Burt JE, Barber GM (1996) Elementary statistics for geographers, 2nd edn. Guilford, New York

Canadian Mortgage and Housing Corporation [CMHC] (2021) Historical average rents by bedroom type. https://www03.cmhc-schl.gc.ca/hmip-pimh/en/TableMapChart/

Canadian Rental Housing Index (2016) Rental housing data. http://rentalhousingindex.ca/en/

CBRE (2017) An overview of Airbnb and the hotel sector in Canada http://new.hotelassociation.ca/wp-content/uploads/2018/06/Full-Report.pdf

Cleave E, Easton M, Arku G (2020) Spatial analysis and evidence-based economic development planning: a case study on manufacturing clustering in southern Ontario. In: Vaz E (ed) Regional intelligence: spatial analysis and anthropogenic regional challenges in the digital age. Springer International, Cham, pp 121–144

CMHC (2019) Housing market information portal. CMHC, Toronto. https://www03.cmhc-schl.gc.ca/hmip-pimh/en#TableMapChart/2270/3/Toronto

Combs J, Kerrigan D, Wachsmuth D (2020) Short-term rentals in Canada: uneven growth, uneven impacts. Can J Urban Res 29(1):119–134

Deane S (2021) 2021 Airbnb statistics: usage, demographics, and revenue growth. Retrieved from https://www.stratosjets.com/blog/airbnb-statistics/

Deboosere R, Kerrigan D, Wachsmuth D, El-Geneidy A (2019) Location, location and professionalization: a multilevel hedonic analysis of Airbnb listing prices and revenue. Reg Stud Reg Sci 6(1):143–156

FairBnB (2020) Increasing Toronto's vacancy rate. FAirbnb Coalition:1–8

Frenken K, Meelen T, Arets M, Van de Glind P (2015) Smarter regulation for the sharing economy. Retrieved from https://www.theguardian.com/science/political-science/2015/may/20/smarter-regulation-for-the-sharing-economy

Gallichan-Lowe S (2020) Toronto introduces new rules for AirBnB, other short-term rental platforms. Retrieved from: https://toronto.citynews.ca/2020/08/25/toronto-introduces-new-rules-for-airbnb-other-short-term-rental-platforms/

Gibbs C, Guttentag D, Gretzel U, Morton J, Goodwill A (2017) Pricing in the sharing economy: a hedonic pricing model applied to Airbnb listings. J Travel Tour Mark 35(1):46–56

Grisdale S (2018) What Airbnb is doing to Toronto's rental apartment market. Spacing Toronto. http://spacing.ca/toronto/2018/04/25/Airbnb-torontos-rental-apartment-market/

Gutiérrez J, García-Palomares JC, Romanillos G et al (2017) The eruption of Airbnb in tourist cities: comparing spatial patterns of hotels and peer-to-peer accommodation in Barcelona. Tour Manag 62:278–291

Guttentag D, Smith S, Potwarka L, Havitz M (2018) Why tourists choose Airbnb: a motivation-based segmentation study. J Travel Res 57(3):342–359

Heisz A LaRochelle-Côté S (2005) Work and commuting in census metropolitan areas, 1996–2001. Trends and conditions in census metropolitan areas. Analytical studies branch. Catalogue no. 89-613-MIE2005007. Statistics Canada, Ottawa

Horn K, Merante M (2017) Is home sharing driving up rents? Evidence from Airbnb in Boston. J Hous Econ 38(Complete):14–24

Jamasi Z (2017) Regulating Airbnb and the short-term rental market: an overview of North American regulatory frameworks
Krause A, Aschwanden G (2020) To Airbnb? Factors impacting short-term leasing preference. J Real Estate Res 42(2):261–284
Lima V (2019) Towards an understanding of the regional impact of Airbnb in Ireland. Reg Stud Reg Sci 6(1):78–91
Local Planning Appeal Tribunal (2019) PL180082: decision delivered by S. Tousaw and order of the tribunal. Local Planning Appeal Tribunal
Mermet A-C (2017) Airbnb and tourism gentrification: critical insights from the exploratory analysis of the 'Airbnb syndrome' in Reykjavik. In: Gravari-Barbas M, Guinand S (eds) Tourism and gentrification in contemporary metropolises. Routledge, New York, pp 52–74
Nelson JP (2010) Valuing rural recreation amenities: hedonic prices for vacation rental houses at Deep Creek Lake, Maryland. Agric Resour Econ Rev 39(3):485–504
Oskam J, Boswijk A (2016) Airbnb: the future of networked hospitality businesses. J Tour Futures 2(1):22–42
Rifkin J (2014) The zero marginal cost society: the internet of things, The collaborative commons, and the eclipse of capitalism. Palgrave Macmillan, London
Rogerson PA (2010) Statistical methods for geography, 3rd edn. Sage, Washington
Shabrina Z, Arcaute E, Batty M (2021) Airbnb and its potential impact on the London housing market. Urban Stud 59(1):197–221. https://doi.org/10.1177/0042098020970865
Sheppard S, Udell A (2018) Do Airbnb properties affect house prices? Williams University Working Paper. Available at: https://pdfs.semanticscholar.org/c41a/555192f8b8a63413cfd461f4295b95123e6f.pdf?_ga=2.259932618.15608359.1596668421-1815735046.1596668421
Sovani A, Jayawardena C (2017) How should Canadian tourism embrace the disruption caused by the sharing economy? Worldw Hosp Tour Themes 9(4):464–470
Toronto Regional Real Estate Board [TRREB] (2021) Historical statistics. https://trreb.ca/files/market-stats/market-watch/historic.pdf
Tsai H, Huang WJ, Li Y (2016) The impact of tourism resources on tourism real estate value. Asia Pac J Tour Res 21(10):1114–1125
Wachsmuth D, Weisler A (2018) Airbnb and the rent gap: gentrification through the sharing economy. Environ Plan A: Econ Space 50(6):1147–1170
Wachsmuth D, Kerrigan D, Chaney D, Shillolo A (2017) Short-term cities: Airbnb's impact on Canadian housing markets. Urban Politics and Governance Research Group
Wieditz T (2017) Squeezed out: Airbnb's commercialization of home-sharing in toronto. Retrieved from http://fairbnb.ca/wp-content/uploads/2017/09/Fairbnb_Report_Feb_29.pdf
Woodward J (2020) Data scan finds more than half of Vancouver Airbnbs are breaking city rules. https://bc.ctvnews.ca/data-scan-finds-more-than-half-of-vancouver-Airbnbs-are-breaking-city-rules-1.5203296
Yrigoy I (2019) Rent gap reloaded: Airbnb and the shift from residential to touristic rental housing in the Palma Old Quarter in Mallorca, Spain. Urban Stud 56(13):2709–2726
Zervas G, Proserpio D, Byers J (2016) The rise of the sharing economy: estimating the impact of Airbnb on the hotel industry. In: Boston University School of Management research paper (2013–16), Cambridge

Chapter 9
Urban Agriculture and the Neighborhood Context: A Geodemographic Approach to the City of Toronto

9.1 Introduction

This chapter presents a study of the demographic characteristics of populations in areas surrounding urban agricultural sites in the City of Toronto, Ontario, Canada. Urban agriculture refers to the growing of produce for consumption within a built-up space, generally within a city or town. This produce can take the form of edible plants and vegetables or flowers. Urban agriculture is becoming an important issue for a number of reasons. From an economic perspective, the average Canadian grocery bill is set to rise significantly over the coming years (Evans 2021), while from a livelihood perspective, food security is becoming an ever-greater concern for the majority of Canadian families. More broadly, climate change poses major risks to the global food supply chain. Individual and national self-sufficiency in this arena thus become ever more important. In addition to this stark reality, urban agricultural and allotment sites provide important spaces for cultural exchange and intergenerational encounters (Kleinschroth & Kowarik 2020; GrowTO 2012). Indeed, as GrowTO (2012, p. 9) argues:

> The economic benefits of urban agriculture can be found at every scale, from the money saved in individual household budgets by growing food, through to the potential cost savings to provincial healthcare budgets as a result ofincreased vegetable consumption and increased social cohesion.

As urbanization in Toronto continues to increase, the rate of development and densification in the urban core will dramatically increase. In response, policies must consider the importance of spaces such as these and seek to incorporate them into developing neighborhoods and target those areas which would most benefit from them.

The literature surrounding allotments and their role in community life is diverse and deals with a wide range of topics from the importance of urban agriculture in achieving the sustainable development goals (Nicholls et al. 2020) to their role in

E. Vaz, *Regional and Urban Change and Geographical Information Systems and Science*, Advances in Geographic Information Science,
https://doi.org/10.1007/978-3-031-24731-6_9

the extended health of older populations (Dennis et al. 2020). Much of this literature deals with qualitative data surrounding site users and their specific beliefs and attitudes toward cultural, societal, and environmental trends (see Christensen et al. 2019). In contrast, Håkansson (2021) argues that the existing literature places an overemphasis on determining the demographic elements of site users which lead to "successful" urban agricultural sites while neglecting the analysis of those sites which fail. Rather than simply seeking to determine those demographic factors which best support the persistence of urban agriculture spaces, Håkansson (2021) argues that no one demographic factor can truly explain the persistence of these spaces. Additionally, the array of demographic, socio-economic, and physical factors that lead to the creation and maintenance of these spaces cannot be compressed at the neighborhood level and exported to other sites. Despite this, however, Håkansson (2021) maintains that there is inherent value in understanding the demographics of the neighborhoods which surround these sites at the municipal scale.

Building on Christensen et al. (2019), Dennis et al. (2020), Håkansson (2021), and Singleton and Longley (2015), this research takes the perspective that demographic factors and patterns can provide insight to municipalities seeking to improve urban agriculture and social cohesion. While it is true that no model can completely capture the diversity of factors that underpin the success of these sites on a national or global scale, there is room to explore a series of interacting models which may be used to inform policy decisions on a municipal scale. After all, some reductionism is necessary when crafting municipal policy, nor can every possible factor be assessed when making those policy decisions.

Within this context of increased need and use of urban agriculture, this approach presented in this chapter sets out to create a broad demographic classification of dissemination areas (DAs) in Toronto. Building on the methodology first presented by Singleton and Longley (2015), this study will use a tiered k-means cluster analysis to separate areas along a variety of demographic lines. Following this, the study will assess the dominant demographic characteristics of the DAs which surround both privately and publicly run urban agricultural sites throughout the City of Toronto. In summary, the objectives of this study are:

- RO1: to develop a demographic segmentation of potential urban agriculture users through clustering analysis.
- RO2: identify who has access (based on proximity) to different types of urban agricultural sites in Toronto.

The study advances two main hypotheses about the demographic characteristics of urban agricultural sites in Toronto. Firstly, similar to the European context (see Håkansson 2019, 2021), city-run urban agricultural sites will exist primarily in lower-income neighborhoods with lower overall densities. Secondly, in contrast to this, privately run urban agricultural sites are expected to be found in both high-density and higher-income areas. The study hypothesizes that these areas will be dominated by younger knowledge economy workers and students—in essence, a population that would formerly have been considered as evidence of early-stage gentrification.

9.2 About the City of Toronto

The City of Toronto, Ontario, is the study area for the analysis presented in this chapter. It represented one of the most diverse and fastest growing urban centers in North America (Patton 2021). The city is the most populous in Canada with a diverse population of approximately 2.93 million inhabitants in the metropolitan area.

Within Toronto, there is acknowledgment of the need for urban agriculture systems. The Toronto and Region Conservation Authority (a community-based natural resource management agency) operates approximately 400 ha of agricultural land within the Greater Toronto Area with a goal of developing local and sustainable food systems. The TRCA notes the importance of urban agriculture as "providing space for people and organizations to grow and harvest fresh food is our way of bringing communities together and feeding both bodies and minds" (Toronto Conservation Authority 2022). The City of Toronto also considers urban agriculture as a policy issue. The city's official plan highlights urban agriculture as important "for creating beautiful, healthy and active cities and for engaging diverse communities" (as quoted in GrowTO 2012, p. 3). Indeed, GrowTO's (2012) *An Urban Agriculture Action Plan* represented a collaboration of city government, nonprofit organizations, and private stakeholders to "highlight the economic and social development opportunities that urban agriculture can bring to communities and neighbourhoods" (p. 1) and developed a pathway to enhancing urban agriculture in the city. This includes developing greater awareness of local agriculture programs, providing training, opportunity, and access in an equitable manner.

Despite the perceived need for urban agriculture infrastructure in Toronto, the city currently has only 12 city-administered urban agricultural sites, as shown in Fig. 9.1. In contrast to this, there are 173 publicly accessible sites which fall into the privately run category, as shown in Fig. 9.2 which summarizes all sites (including publicly run) within the city. This category includes sites run by neighborhood or nonprofit associations.

For the purposes of this study, only those sites which can be deemed publicly accessible or rooftop sites are included in the analysis. Thus, the actual number of urban agricultural sites in Toronto is likely significantly higher given the proliferation of urban growers in recent years. However, while sites which are privately run may be publicly accessible, there are often barriers to entry in the form of memberships or annual fees. Future city development policies should take into account the need for self-sustaining urban agriculture and the growing importance of outdoor community hub spaces for marginalized populations—both of which functions allotment sites serve.

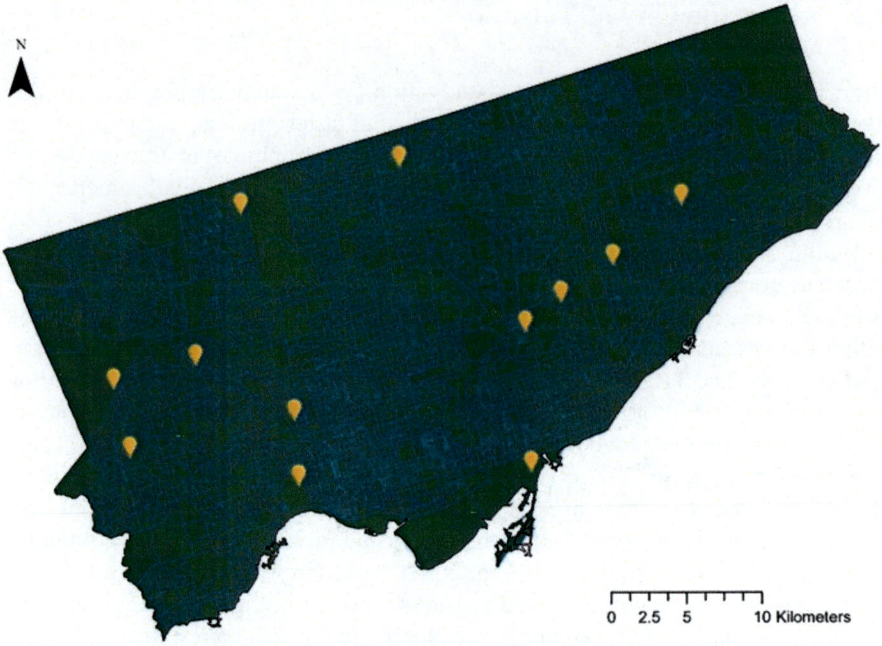

Fig. 9.1 City-run urban agricultural sites in the city of Toronto, Ontario

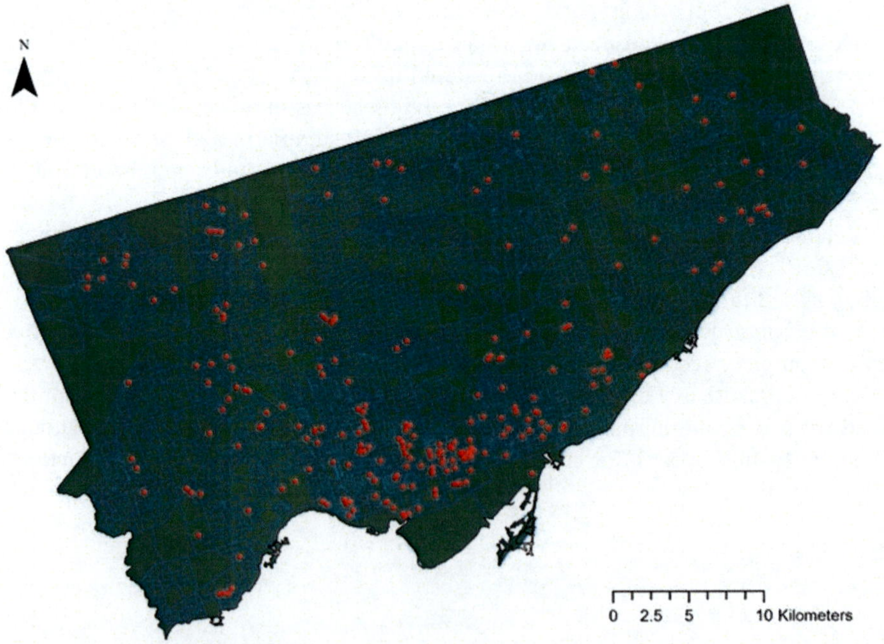

Fig. 9.2 Publicly accessible urban agricultural projects in the city of Toronto, Ontario

9.3 Background

Conceptually, this chapter centers its focus on the relationship between urban agriculture and gentrification. Håkansson (2019, 2021), also building off of Singleton and Longley (2015), explored the prevalence and persistence of such sites in the British context, analyzed over 1500 sites in the City of London (in contrast, there are 262 sites in Toronto), drawing three key conclusions: (1) lower-income neighborhoods with a high degree of non-native-born individuals are receptive to community-led sustainability initiatives such as urban agriculture; (2) gentrifying neighborhoods or those in early-stage gentrification support the creation and persistence of similar sustainability initiatives; and (3) late-stage gentrified neighborhoods are not conducive to the creation or persistence of such projects.

When considering allotment sites and green infrastructure broadly, the neighborhood is only one of several factors which play into the persistence of such projects. Christensen et al. (2019) provide an analysis of several other factors at play within these projects in the Danish context, identifying the formation of social capital through integration, with an allotment garden site as the locus for this integration and capital formation. Christensen et al. (2019) report that while users of the allotment site generally believed that the garden was a place of positive exchange and broad diversity, the demographics of the surrounding neighborhood did not map well onto the demographics of the site users. The area around the site was highly diverse and largely immigrant-occupied, while the garden users tended to be Danish, or Danish born (Christensen et al. 2019).

9.3.1 Green Infrastructure and Urban Processes

In the broader urban context, there is much greater uptake of urban agriculture in European cities, compared to Toronto and its North American counterparts. Potential explanations of this slow uptake of agricultural infrastructure may lie in the layout of cities. North American cities have relatively small urban cores with the dominant type of housing being single-family detached homes outside of these cores (Moos et al. 2017). This housing type is often combined with relatively large lot sizes, especially outside of the urban core. As Håkansson (2019) notes, this typology often leads to less reliance on public green infrastructure as residents generally have sufficient space to meet their needs. However, those populations that are unable to afford these properties are increasingly left behind by municipalities. In fact, as cities like Toronto densify at ever-increasing rates, the importance of green infrastructure projects is set to re-emerge as a dominant consideration in the urban landscape.

Green infrastructure projects (which include urban agriculture) have become explicitly linked with gentrification in the literature, where it is often defined as "ecological gentrification." Harper (2020) points out that green infrastructure projects are increasingly employed by developers to attract specific populations into an area. As climate anxiety is increasingly on the rise, the middle class seeks out developments which market themselves as "climate neutral" or "net zero." These types of developments rely upon "greening" of sites as a marketing strategy to entice younger populations (Harper 2020). The use of sustainability as a marketing tool relying upon the inclusion of green infrastructure in new developments points to its importance in the cultural mind of the West. This includes the middle classes who traditionally represent gentrifiers in the extant literature.

Rice et al. (2020) expand upon the concept of ecological gentrification by including lifestyle choices, such as mode-of-commute, as an alternative indicator of gentrification, termed "carbon gentrification." Young professionals who are attracted to the type of development described more likely to commute via foot or bike, thus needing to be in central urban areas (see Harper 2020; Rice et al. 2020). Again, these populations are also attracted to developments which include green infrastructure as indicators of sustainability, whether real or imaged. This would largely support Håkansson's (2019, 2021) assertions that that early-stage gentrifiers are receptive to urban agricultural projects, as these are a primary component of green infrastructure as a sustainability indicator.

Much of the early literature on gentrification which utilized the stage approach to analyze both gentrification's impacts and processes relied upon it being a local process affected heavily by local factors (Gale 1979; Kerstein 1990). However, with the rise of global capitalism, there has been an accompanying rise in a form of entrepreneurial global urbanism. This type of urbanism often sees municipalities behaving as private actors attempting to attract global investment to the city. Smith (2002) explicitly ties gentrification to this form of globalized urbanism, positing that policies which actively support gentrification in urban cores are being utilized by municipal actors to attract both talent and industry to cities. This is somewhat at odds with the localized vision of a stage model of gentrification. The stage model, as espoused by Kerstein (1990; see this paper for a full review of contributors to the stage model up until this point) and Gale (1979), envisions gentrification as being carried out by local private actors. The first wave of gentrifiers in this model are those who are risk-oblivious, populating an area where heritage buildings are in disrepair and amenities are not necessarily abundant. These first-wave colonizers give way, in the final stage, to the risk-averse. Higher-income middle-class inhabitants are drawn to the area once there is a degree of cultural homogeneity in the neighborhood. This version of gentrification sees it as a gradual process whereby the original inhabitants are pushed out by waves of newcomers (who themselves may be pushed out by a later wave).

In contradiction to this, gentrification in the global city becomes municipality sponsored. The city attempts to volt the gradual process outlined above and by dint of policy and zoning choices moves to late-stage gentrification as outlined above. This vision of gentrification applies well to the City of Toronto, where increasing densification and property values in the entirety of the census metropolitan area would see a sidelining of the stage model of gentrification in favor of a municipality-sponsored jump to late-stage gentrification. Smith (2002) makes the point that the rise of global capitalism has coincided with the rise of state-sponsored gentrification. Smith finds that cities which exist as nodes in the global economy pursue municipal policies which actively seek to encourage a new wave of gentrification.

In step with this model of municipal support as a major driver of gentrification is the concept of youthification (Moos 2016). This contrasts Håkansson (2019, 2021), as well as other descriptions of gentrification, arguing that the traditional use of a high percentage of young adults in an area as an indicator of gentrification is flawed. This is because youth exist as a separate dimension of socio-spatial differentiation (Moos et al. 2017). The underlying argument is that the process of gentrification is, in a sense, complete in the majority of dense, global city cores and as a result there is a shift in capital investment patterns and changes in life cycle where "early-stage gentrifiers" do not necessarily give way to the later stages of the model. They remain in situ as they progress in their careers, seeking out the urban core in preference to those inner-city suburbs so favored by earlier gentrifiers (Moos et al. 2017). Indeed, one of the central features of youthification is the concentration of the young along transit lines. This indicates yet another life cycle change as compared to earlier models dealing with gentrification and plays well with the concept of carbon gentrification outlined by Rice et al. (2020).

In a parallel process, Lehrer and Wieditz (2009) discuss a process of condoification as urban cores begin to increasingly build up in search of space. These spaces are inhabited largely by the young and the very elderly. The young are increasingly choosing to live in these dense urban spaces, frequently chosen for their proximity to work or amenities. This spatial stratification of age groups suggests a more expansive view of neighborhood characteristics is needed. This would justify the inclusion of the presence of younger age groups as an indicator of interest in sustainability initiatives such as urban agriculture, whether or not age was explicitly tied to the forms of gentrification (i.e., expanding the scope from previous gentrification research).

9.3.2 Practical Considerations: The Need for Analysis

Based on this, policy could be developed to understand which neighborhoods are most receptive and fitted to exist as seed sites in a city. However, it is difficult to reduce the complex array of factors at play in the persistence of such projects to a neat demographic classification (Håkansson 2019). This view of urban spaces as places of great complexity is articulated by Fuenfschilling (2019, p. 221) where "urban spaces are characterized by a high degree of complexity and interrelation of various actors and sectors, by a high concentration of resources…or by high diversity of actors in regard to socio-economic backgrounds or education." Loorbach et al. (2017) further underline that the complexities exhibited by urban spaces make categorization of any type difficult, and its replication on an international scale even more so. Fuenfschilling (2017) continues this trend by exploring the ways that the attributes initially mentioned by Loorbach (2017) could impact the creation and persistence of such sustainability projects in either a positive or negative way depending upon the local context.

Kronsell et al. (2018), however, explore the idea of how a municipality may play a key role in supporting the use and growth of sustainable experiments and urban agriculture. They outline three main roles that a municipality can play: promoter, enabler, or partner (Kronsell et al. 2018). These roles, while useful for understanding sites that are not municipally run, do not fit perfectly onto projects which are conceived of and funded by the city in their totality. Such is the case in Toronto, where a variety of the sites under study are exclusively city-run. This offers an important insight into the role of the municipality across a broader landscape of policy and interaction with grassroots or privately run organizations that is worthy of consideration.

9.4 Methodology

Methodologically, this research utilizes Singleton and Longley (2015) as the "jumping off point" for the analysis of Toronto. The analytical framework has also been applied by Håkansson (2021); however, both studies were situated in London, England. This study's current classification system was built along similar lines to allow for comparability between the studies to understand the major differences between the British and North American contexts. Singleton and Longley (2015) set out to create a geodemographic classification of the London census output areas (OA) based upon a series of variables controlled for intercorrelation. The purpose of this method was to demonstrate that, while large-scale classification systems have their uses, classifications at the local scale can reveal more detailed demographic trends specific to the area under study. London was selected as the study area since it was considered to be the only "truly global city" in the United Kingdom at the time of the study (Singleton and Longley 2015). Additionally, there is a flattening of regional differences when engaging with demographic trends, and there is a need for regional classifications to truly understand the local context (i.e., the city level), finding that a local classification produced distinct trends which were not apparent in the 2011 OA classification at the national level.

9.4.1 Data

The data used for this study comes from two principal sources: Environics Analytics and the Toronto Urban Growers. Environics Analytics DemoStat 2021 projections were used for all the variables used in the geodemographic classification of Toronto. Variable choice was influenced by the relevant literature on urban agriculture (Håkansson 2019, 2021; Singleton and Longley 2015) and on gentrification (Rice et al. 2020) and its "spin-offs" of condo-ification (Lehrer and Weiditz 2017) and youthification (Moos 2016). Broadly, the selected variables can be broken down into several broad groups: age-based, ethnicity-based, mode-of-commute, dwelling type, family structure, employment, and educational attainment. In total they encompass 75 Canadian census variables collected at the DA level (as rates or averages). Each variable was analyzed for both skewness and kurtosis (using Fisher's skewness and kurtosis metrics) through a Pandas python script and was transformed (i.e., Log10) as required.

The data on urban agriculture and green infrastructure projects more broadly in the City of Toronto posed a significant initial problem for the study. The City of Toronto does not publish data on anything but green roofs in the downtown core, meaning that to acquire the data, significant and time-intensive scraping of the data points from Toronto Urban Grower's (TUG) website is required. Data was winnowed down to include only those projects which might be deemed accessible to the public. TUG breaks down projects into several categories: Toronto park's community gardens, public housing community gardens, other community gardens, allotment gardens, school and children's gardens, producers, other urban agri-businesses, organizations, rooftop gardens, greenhouses, beehives, orchards, and others. Of this exhaustive list, the three types of community gardens, the allotments, and the rooftop gardens were retained ($n = 262$), while all others were discarded as having barriers to access or falling outside the definition of urban agriculture.

9.4.2 Data Reduction

The segmentation analysis used for this study was conducted in two steps: first, a variable selection/reduction procedure was conducted; and second, a clustering analysis was employed. As noted, the initial variable list encompassed 71 variables that were broadly similar to those used by Singleton and Longley (2015). The first step of the data reduction which was an exploratory K-means clustering analysis was performed. The initial analysis used eight clusters to match that of Singleton and Longley's (2015). This initial clustering was then examined via a discriminant analysis to assess variable and group performance. Overall, the initial discriminant analysis suggested that between-group covariance was high and so does within-group covariance. The canonical function tables supported these results and showed several distinct outliers. Additionally, the territorial map pointed to confusion between several of the groups, despite the initially high levels of between-group covariance. The structure matrix indicated that over half of the initial 71 variables were causing issues in the analysis.

Based upon these results, a stepwise discriminant analysis was then run to determine which variables were contributing most to model fit. The stepwise discriminant analysis winnowed the initial list of 71 variables down to 39. Most variables removed fell into three groupings. The first grouping was comprised of age variables. Interestingly, the stepwise discriminant discounted most of the age variables dealing with post-retirement age populations. The second grouping was minority status. Here the stepwise discriminant process removed most of the minority status variables. This likely was the result of a high number of DAs with 0% minority status for a large number of ethnic groups. The third and final grouping was industry. Here the stepwise discriminant removed almost all participation by industry variables, leaving only four from the initial list of participation by industry variables.

9.4.3 Segmentation

Following data reduction, a two-step cluster analysis was run to ensure an optimal number of clusters were chosen for the final analysis. Using Schwarz's Bayesian information criterion, the silhouette measure of cohesion and separation assessed a five-cluster solution as suitable. Based on the group selection, K-means clustering analysis with five final clusters was conducted. These five clusters represent supergroups, representing the upper tier of analysis.

The five "supergroups", formed in the clustering analysis, "Established Suburban Professionals," "Aging Suburban Retirees," "The Struggling Middle," "Young Gentrifiers," and "Educated Knowledge Economy Workers," provide a strong and cohesive categorization of the City of Toronto. The canonical discriminant function (see Fig. 9.3) shows that the groups have high levels of within-group cohesion as well as between-group differentiation. This in turn means that any conclusions drawn from the distribution of groups throughout the city will have a high degree of

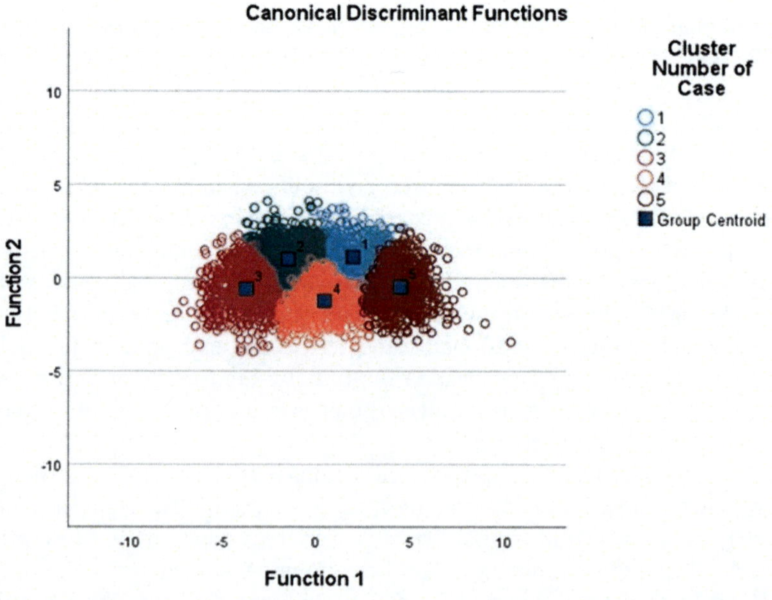

Fig. 9.3 Canonical discriminant function

significance. In addition to this, green infrastructure and urban agricultural projects in each of the clusters differ greatly in numbers. This further helps to solidify any conclusions drawn from the data.

K-means was used again within each supergroup cluster to create the secondary grouping needed for a multitier analysis. Given the granularity of the data being investigated, the decision was taken to use a set number of clusters for this second step clustering. Singleton and Longley (2015) suggest three clusters for the second stage clustering. This yields a total of 15 clusters at the second tier of clustering. Each supergroup's final cluster center was analyzed by comparing the centers to overall variable means and standard deviations, in order to assess both the importance of the variable and which variables drove differentiation. A similar process was followed with the subgroups' final cluster centers. Finally, the count of the urban infrastructure projects within each supergroup and "subgroup" was conducted to understand the distribution of allocation sites across different types of neighborhoods.

9.5 Cluster Descriptions

The data reduction and clustering analysis identified five supergroups which each exhibited unique trends. These groups were "Established Suburban Professionals," "Aging Suburban Retirees," "The Struggling Middle," "Young Gentrifiers," and "Educated Knowledge Economy Workers" (summarized in Table 9.1 and Figs. 9.4 and 9.5). Descriptions of each of the five supergroups and their subgroups follow.

Established Suburban Professionals This group (592 DAs) presented with higher numbers of young children (between the ages of 10 and 14). Overall, the group had average to high educational attainment and was predominantly employed in the professional, science, or technological fields; however, there were other groups with higher employment in these areas. This group primarily drove to work, either as a driver or as a passenger in a vehicle. The group favored less dense housing, with a low score on the variable for the 5+ storey apartment building structure type. This was relatively unsurprising given that the spatial distribution of the group puts it largely outside of the center of Toronto.

Within this group, three subgroups were produced by the secondary k-means analysis. These three groups separated largely along age, employment, and style of living lines. The first subgroup, New Entrants to Suburban Living, had some educational attainment and was employed in the info and culture and food and accommodations sectors mainly. They tended to be slightly younger as represented by the 25–29 variable and were more likely to commute in an eco-friendly way. They also tended to live in slightly denser housing types. The second subgroup, Standard Suburbanites, had lower educational attainment overall, tended to travel to work by car, and live in non-dense housing types. The third subgroup, Established Wealthy Suburbanites, represented the group with the highest educational attainment in this

Table 9.1 Summary of market segmentation of supergroups and agricultural site distribution

Supergroup	Subgroup	Dissemination areas	Agricultural sites	Index score
1. Established suburban professionals	New entrants to suburban living	221	9	
	Standard suburbanites	200	6	
	Established wealthy suburbanites	171	5	
	Total	**592 (16%)**	**20 (8%)**	**50**
2. Aging suburban retirees	Blue collar families	277	13	
	Marginalized youth	375	16	
	Traditional suburban retirees	267	4	
	Total	**919 (25%)**	**33 (13%)**	**52**
3. The struggling middle	Outliers	7	0	
	Struggling retirees	348	35	
	Working young families	512	55	
	Total	**867 (23%)**	**90 (34%)**	**147**
4. Young Gentrifiers	The risk oblivious	221	16	
	Grown in place	180	24	
	Retirees hanging on	236	31	
	Total	**637 (17%)**	**71 (27%)**	**158**
5. Educated knowledge economy workers	Successful urban young	169	23	
	Established Inner City suburb families	248	5	
	Artistic, eco-conscious wealthy	270	20	
	Total	**687 (19%)**	**48 (18%)**	**94**

Note: Index score = 100 * (dissemination area %/agricultural sites %)

set. Those living in DAs assigned to this subgroup were generally employed primarily in professional, science, or technological industries and primarily drove to work.

Aging Suburban Retirees This group (919 DAs) was spatially distributed toward the Eastern and Western edges of the city (see Fig. 9.4). They presented with an overrepresentation of populations either close to retirement age or in retirement. They were unlikely to have children at home and had relatively low educational attainment. This group also tended to drive to work and largely did not live in high-density dwelling types.

This supergroup was again split into three subgroups. The first of these subgroups, Blue Collar Families, had more young children (10–14 years old), tended to have apprenticeships or trades certificates of some type, and primarily used cars, either as a passenger or as a driver, to commute to work. The second subgroup, Marginalized Youth, had the highest proportion of young adults among the Aging Suburban Retiree supergroup. Those living in neighborhoods assigned to this

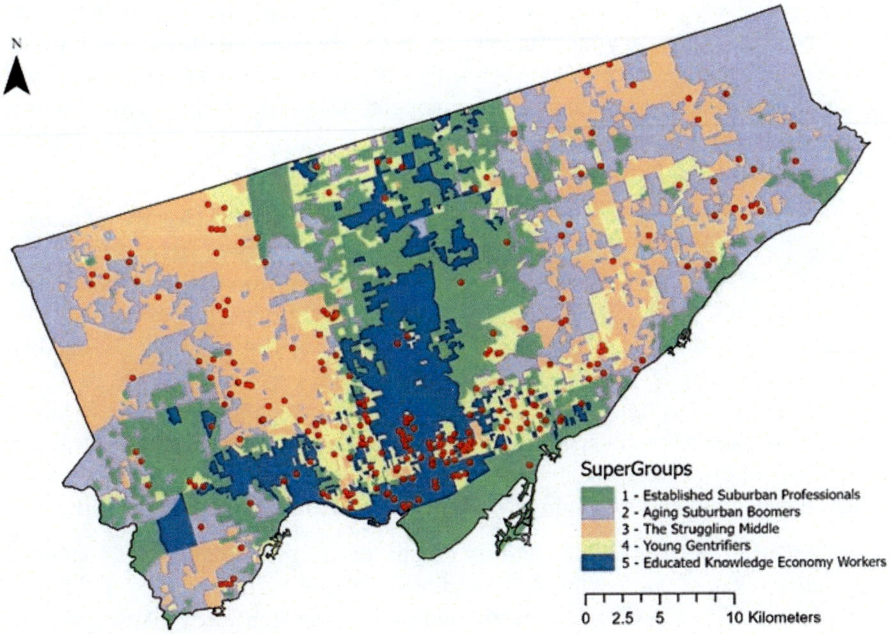

Fig. 9.4 Distribution of supergroups

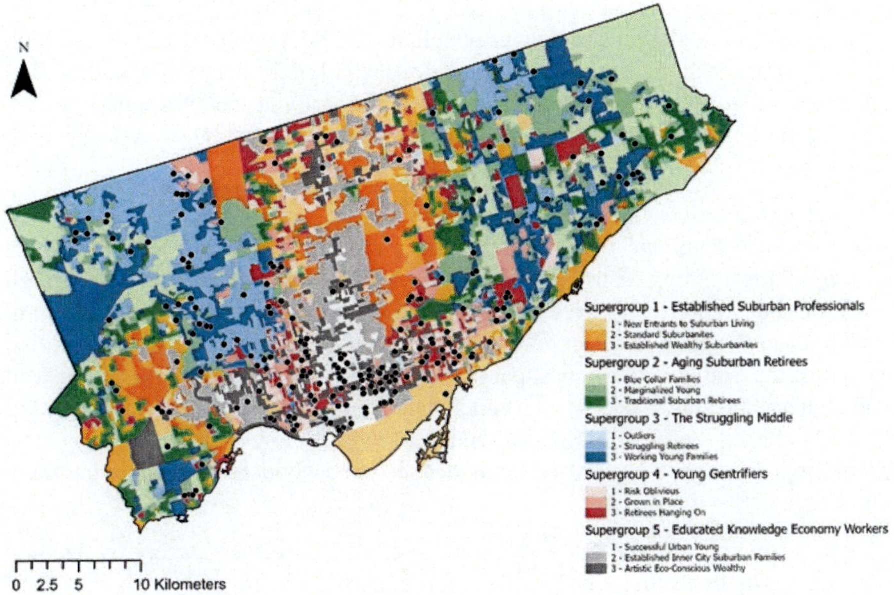

Fig. 9.5 Distribution of subgroups throughout supergroups

cluster generally drove less and lived in more dense structure types. They tended to have lower rates of employment in general, but those working over-indexed in the food and accommodation services sector. The third subgroup, Traditional Suburban Retirees, had high proportions of aging populations—the 55–69-year-old cohort were most heavily represented in this subgroup. This group was characterized by families without children in the home and did not live in dense areas.

The Struggling Middle This group (867 DAs) was distributed slightly more centrally in Toronto than the "Aging Suburban Retirees" grouping, largely occupying an area to the West of Toronto in a North-South swathe, stopping just short of the downtown area (see Fig. 9.4). Compared to other supergroups, this cluster had significantly lower educational attainment in most higher education categories and had higher than average levels of apprenticeship or trades certificates and high school diplomas as the education level. This group tended to be employed in the food and accommodation sectors and generally drove, mainly as a passenger, to work. They also exhibited low levels of employment compared to other clusters and low levels of families without children at home (i.e., they tended to have younger families).

When this group was divided into subgroups, one of the subgroupings contained only outliers. These outliers had 0 values for all or most variables. They will not be discussed in this analysis as the likelihood is that they represent disused land or properties under development. The first subgroup which contained usable results, Struggling Retirees, had a generally older population with very low overall educational attainment. This group also tended to live in semidetached housing and drove to work, either as a driver or as a passenger. The second usable subgroup, Working Young Families, had a much younger population and the family structure had high rates of children in the home. They tended to have higher degrees of educational attainment than subgroup 2; however, this was primarily at the high school or college levels. They also lived primarily in apartments and had high overall levels of employment.

Young Gentrifiers Areas of Toronto assigned to this group (637 DAs) were found to be spatially dominant in the downtown core, primarily the Western end of the core but with a strong distribution in the East side as well. This cluster had high levels of young adults who were primarily employed in the information and cultural sectors or in the food and accommodation sectors. Those living in areas identified as part of the Young Gentrifiers supergroup tended not to drive to work and instead had high proportions of people who either walked or took public transit. As expected in the center of the city, this group tended to live in more dense structure types. Surprisingly though, they favored semidetached or row houses over large apartment buildings.

Within this cluster, the three subgroups were distinguished largely by age and education. The first subgroup, The Risk Oblivious, had significantly higher levels of young adults and overall higher educational attainment. This group tended to work in the information and cultural sectors and preferred to bike or walk to work. The second subgroup, Grown in Place, had slightly higher levels of middle to late middle-aged populations. Their educational attainment was in more practical areas such as college diplomas and apprenticeship or trades certificates. They were also more likely to have children in the home than either of the other subgroups. The third subgroup, Retirees Hanging On, had a high proportion of retirees. This group had lower levels of employment and had the lowest educational attainment of the supergroup.

Educated Knowledge Economy Workers Finally, as Figs. 9.4 and 9.5 summarize, this supergroup (687 DAs) was generally located along a North-South axis in the center of Toronto. Geographically, this area can be characterized by its proximity to the Don Valley Parkway, one of the largest green space corridors in the Toronto area. Some of Toronto's most affluent neighborhoods, such as Rosedale, are covered by this grouping. This cluster exhibited high employment in the professional, scientific, and technical sectors, as well as relatively high levels of employment in the information and cultural sectors. This cluster also had significantly higher levels of educational attainment, with the highest levels of populations with bachelor's degrees or above bachelor's degrees. Those living in areas identified as part of the *Educated Knowledge Economy Workers* group tended to commute to work by foot or by public transit and had higher levels of large apartment buildings as the dwelling type.

The first subgroup, the Successful Urban Young, had the highest levels of young adults, the highest educational attainment, and worked primarily in the professional, scientific, and technical sectors. They also tended to walk to work and live in apartments without children. The second subgroup, the Established Inner City Suburb Families, had high levels of both 10–14 years old and retirees (60–69 years old), with lower overall levels of education. This group tended to drive to work and had low levels of large apartments and semidetached housing. The final subgroup, Artistic, Eco-Conscious Wealthy, was primarily employed in the information and cultural sectors and tended to bike or take public transit to work. They also had a high proportion of semidetached houses in this subgroup.

9.6 Analysis of Urban Agriculture Distribution

The distribution of green infrastructure projects and urban agricultural projects throughout these five supergroups and their associated 15 subgroups is summarized in Table 9.1. *Established Suburban Professionals* has the lowest number of projects at 20, followed closely by *Aging Suburban Retirees* at 33. As noted in Table 9.1, both of these supergroups were under-indexed (scores of 50 and 52, respectively), meaning that these segments had fewer agricultural sites than expected based on their prevalence in Toronto (based on the number of DAs assigned to each group). *Educated Knowledge Economy Workers*, containing 48 sites, strikes a middle ground (index score of 94). Of the final two supergroups identified, Young Gentrifiers had the highest number of sites (71) and an index score of 158, while The Struggling

Middle had 90 sites and an index score of 147. Notably, both supergroups had a larger number of sites than expected based on the number of member DAs. Overall, the results indicate some strong spatial patterns to the distribution of green infrastructure and urban agricultural projects in the City of Toronto. This analysis reveals a broad agreement with several theories outlined in the literature as well as supporting the initial hypotheses of the study. The two broad initial hypotheses of this study were firstly, that urban agricultural sites in Toronto would exist primarily in lower-income neighborhoods with lower overall densities and secondly, that these sites will also exist in areas dominated by younger knowledge economy workers and students.

The Struggling Middle As noted, the largest number of agricultural projects are found in DAs occupied by *The Struggling Middle* (Table 9.1 summarized the distribution of agricultural projects in Toronto across the segments developed through the cluster analysis). This cluster contains 90 individual projects which comprise 34.4% of the total number of projects within the city (this group comprises 23% of DAs in the study area). While the analysis did not include any variables pertaining to income or the assumed wealth of a DA, the final model included "% employed" as a variable. This employment variable allows some flexibility when assessing the wealth of an area. Given that the overall educational attainment of this cluster was low, the low values for employment in this cluster suggest that the overall wealth of the cluster is likely to be lower than the other four.

This in turn points to this cluster holding a significant proportion of low-income and/or marginalized households. Therefore, the significant amount of green infrastructure and urban agricultural projects in the cluster agrees with the initial hypothesis that lower-income and/or marginalized neighborhoods would have a higher proportion of such projects. In addition to this, it bears out that a comparison with the London context is worthwhile, as these findings mirror those of Håkansson (2019, 2021).

Within the two subgroups of this cluster, there is also a marked difference in the distribution of such projects. Both subgroups have high levels of projects when compared to the other cluster subgroups; however, one has a notably higher level. The first subgroup, which is termed Struggling Retirees, contains 35 of the 90 projects. As mentioned in the previous chapter, this group contains an older population with low educational attainment who live primarily in row houses. This points to the importance of these projects as spaces for socialization for older populations in lower-income areas. This group lives in a relatively dense style of housing and likely has little space available to them for food or ornamental plant growing. In contrast to this group, the second subgroup from this cluster, termed Working Young Families, contains 55 projects. This represents 20.9% of the total number of projects in Toronto. This group is characterized by lower educational attainment coupled with the presence of children and a predominant structure type of 5+ storey apartment buildings. This indicates that this group likely has little to no access to outdoor space and is more likely to suffer due to inhospitable economic conditions. This subgroup again shows a strong agreement to the initial hypothesis of a higher proportion of green infrastructure projects appearing in lower-income and/or marginalized neighborhoods. However, it is important to note that when examining the numbers of projects per subgroup (and supergroup), there are disparities in the number of cases per grouping. In this case, the number of projects per DA in the Struggling Retirees subgroup is .100, whereas for the Working Young Families, it is .107.

Young Gentrifiers The next most common cluster for agricultural sites to be located was the Young Gentrifiers supergroup. This supergroup contains a total of 71 projects or 27.1% of the overall number of projects (this group represents 17% of DAs in Toronto). In fact, this group had the highest index score (154; see Table 9.1), meaning there were more agricultural sites than expected based on the overall size of the supergroup. This indicates it is the prime segment for urban agricultural uptake.

This group was initially hypothesized to contain a high number of urban agricultural projects, which was heavily supported by the literature. Thus, the presence of a high proportion of the urban agriculture and green infrastructure projects in this supergroup supports the second initial hypothesis of this study. The second hypothesis of the study posited that there would be a concentration of urban agricultural projects in high-density and higher-income areas dominated by younger knowledge economy workers and students. This group is, in essence, early gentrifiers following the stage model (Rice et al. 2020) and the "youthification" class described by Moos (2016).

Regardless of the model employed, this group of young knowledge economy workers living in the city center represents a privileged portion of the population. *Young Gentrifiers* contains all the markers of such a group. It contains high levels of young adults with relatively high levels of educational attainment. These young adults tend to work in the information and culture sectors or the professional, scientific, or technical sectors of the economy. They tend to live in dense developments of either semidetached or larger apartment buildings. In addition to all of this, this population tends to either walk or take public transit to their places of employment. All these aspects of the supergroup represent broad agreement with a variety of the literature outlined in Sect. 9.2 of the study.

The distribution of agricultural projects across the subgroups of this supergroup is relatively even, ranging from 16 to 31. This matches the relatively tight spread of cases across the subgroups (summarized in Table 9.1). The Risk Oblivious subgroup most typically matches the profile of a young urban dweller as outlined above. This group, however, contains the lowest number of urban agricultural projects at 16. The other two subgroups (Grown in Place and Retirees Hanging On) in this cluster demonstrate higher rates of projects with 24 and 31, respectively. Both these groups have slightly higher rates of middle-aged and elderly populations than the first subgroup in the cluster. However, all three subgroups live in structure types which contribute to density, either semidetached housing or in apartment buildings. Given that the location of the majority of DAs in this cluster is in central Toronto, this would theoretically contribute to the higher number of urban agricultural projects in this cluster as these populations are highly space constrained.

Educated Knowledge Economy Workers This group had lower number of urban agricultural projects in the cluster at 48 (18% of sites). This group had 19% of DAs in Toronto classified to it. Based on the existing literature, it is not a surprise that this supergroup would have fewer projects, which this result is consistent with. This supergroup contained 687 total cases with some spread across the subgroups. The Successful Urban Young group encompassed 169 DAs and had with 23 agricultural projects, subgroup 2 contained 248 cases with only 5 of the projects, and subgroup 3 contained 270 cases with 20 of the projects.

Apparent from this breakdown is the overrepresentation of urban agricultural projects within the Successful Urban Young subgroup, characterized primarily by age and housing typology as well as choice of commute type. The group was dominated by the 25–29-year-old age variable combined with the 5+ storey apartment variable and the choice of walking as a commute type. This suggests the group has prioritized distance to employment when choosing dwelling type. Indeed, this group further bears out the views posited by Moos et al. (2017) and Lehrer and Wieditz (2009) on both youthification and condo-ification as this population breaks with the traditional stage model of gentrification.

The second subgroup, Established Inner City Suburb Families, was comprised of more established families in inner-city suburbs. This group tended to drive to work and had low levels of correlation with dense housing types suggesting that they are more likely to live in detached homes, though this variable did not form part of the analysis. This group contained a strikingly low number of urban agricultural projects compared to the other two subgroups. This suggests that the less dense style of living contributes to a lower need for public or semipublic green spaces such as urban agricultural projects. This bears out Håkansson's (2019, 2021) theory that a greater degree of establishment (or potentially a fully gentrified neighborhood) is less likely to contain or be receptive to this type of green infrastructure project.

The final subgroup in this cluster is the more artistic, eco-conscious knowledge economy worker. This group is characterized primarily by their employment in the information and culture sectors and their choice of bike or public transit for commute. This again suggests that the group has prioritized distance to employment as a key factor in location. The subgroup the Artistic, Eco-Conscious Wealthy contains 20 projects, close or on par with the Successful Urban Young (subgroup 1). Overall, these two subgroups, Successful Urban Young and Artistic, Eco-Conscious Wealthy, fit well within the theory of carbon gentrification advanced earlier in this study.

Established Suburban Professionals and *Aging Suburban Retirees* contained the lowest numbers of urban agricultural projects at 20 and 33. The major differentiation between these two groups is age and educational attainment. Both groups exhibited high rates of commute by car to employment and are spatially toward the peripheries of the city—both to the East and to the West. Space is at less of a premium for both these supergroups, and therefore the lack of green infrastructure projects when compared to more central groups is expected. Suburban populations such as these generally favour residences which contribute to lower densities, and lot sizes are, on average, significantly higher than those in the city center.

9.7 Discussion

This research presented in this chapter considered two objectives about the locations of agricultural sites in Toronto, Ontario: (1) to develop a demographic segmentation of potential urban agriculture users through clustering analysis and (2) to identify who has access (based on proximity) to different types of urban agricultural sites in Toronto. Additionally, the two hypotheses advanced by this study are that urban agricultural sites will exist primarily in lower-income neighborhoods with lower overall densities and that these sites will also be found in both high-density and higher-income areas. This secondary hypothesis included a supposition that these areas would be dominated by younger knowledge economy workers and students.

The results of the cluster analysis demonstrated that the population could be segmented (at the DA level) to capture more traditional segments related to gentrification and youthification within Toronto. Additionally, the results of the analysis show that these hypotheses (and answer to research objective two) were generally correct. Despite lacking any variables which deal with the relative wealth of a neighborhood, the overall lower employment and lower educational attainment of The Struggling Middle supergroup provides a strong basis to conclude that this cluster aligns with the lower-income group outlined in the initial hypothesis. This group contained the most urban agricultural projects throughout the city, with the Working Families subgroup containing the majority of these. This supergroup deals primarily with the most economically disadvantaged populations in Toronto, as demonstrated by the relatively low levels of employment for the group. Additionally, this group was not employed in the knowledge economy nor did they exhibit high educational attainment. Thus, while no speculations can be made as to the levels of immigration to these neighborhoods, these DAs tend to contain disadvantaged populations as compared to the remaining supergroups.

The second hypothesis advanced by the study, that projects would also be concentrated in affluent, dense areas, is heavily validated by the data. Both Supergroup 4 and Supergroup 5 fall into this category, given the high degree of educational attainment and employment in both clusters. The entirety of the subgroups in Supergroup 4 follow this typology and subgroups 1 and 3 in Supergroup 5 also adhere to it. Interestingly, the evaluation of this second hypothesis identified several distinct patterns that ran contrary to an interpretation based upon a stage model of gentrification, such as the one advanced by Håkansson (2021). The prevalence of younger, more educated adults in the urban core did initially suggest the possibility of the core being occupied by early-stage gentrifiers. However, the spatial trends exhibited by the supergroups as a whole suggest that a more nuanced process is in play, since this younger group is represented in some of the other clusters at the peripheries of the city.

The results of the study also show strong similarities with the results of Håkansson's (2019, 2021) studies in London, England. This broad agreement between two highly spatially separated locations further supports the study's assertion that Smith's (2002) theory of a global gentrifier class has emerged concretely in the modern globalized city. In effect, regional differences continue to drive some elements of differentiation in globalized urban populations; however, the global economy has produced classes and trends which are identifiable in the populations of "command and control center" cities, which act as hubs in the global network.

Rather than adopting a view consistent with this stage model of gentrification, which envisions gentrification as a gradual process occurring over longer periods of time, a rupture in the urban fabric is necessary in order to understand the prevalent trends in the downtown core. In this instance, it appears that Smith's (2002) model of entrepreneurial urbanism combined with the process of youthification outlined by Moos (2015) and Moos et al. (2017) produces a stronger argument to explain the patterns present in Toronto. The espousal of this model of change does not directly contradict Håkansson's findings, but rather offers an alternative interpretation of the data. Given that the demographic patterns inherent in the placement of urban agricultural projects appear strikingly similar between both London and Toronto, an explanation that draws on more than local gentrification trends is necessary. The entrepreneurial urbanism outlined by Neil Smith, combined with the changing factors affecting location choice among the young, offers a compelling explanation for the trends visible in the segmentation of Toronto. More specifically, early-stage gentrifiers can no longer be seen as a determinant in the presence and persistence of urban agricultural projects. Rather, a youthful population with higher levels of education occupying positions in the global knowledge economy is more indicative of the persistence of such projects.

9.8 Conclusion

This chapter presented an analytical approach—using a tiered k-means classification system on variables chosen by a robust iterative process—to create a methodologically sound framework within which to analyze the distribution of urban agricultural projects across Toronto. This analysis supported the two initial hypotheses of the study which posited that lower-income neighborhoods with a lower overall density would have a high proportion of urban agricultural projects, as would areas with higher incomes and higher densities near the urban core. These findings are in line with those of the European context as outlined by Håkansson (2019, 2021) and further support Smith's (2002) assertion that gentrification follows similar patterns in globalized urban centers.

Based on the findings, there are several potential implications to consider. When crafting municipal policy, there should be a consideration of those local populations which are most likely to be left behind in the global city. Specifically, aging populations on the peripheries of the city are underserved by urban agricultural projects as compared to the vast majority of other populations examined in this study. Future policy should target these vulnerable populations for intervention. In the case of Toronto, policy interventions should focus on micro-scale urban agricultural projects which focus on neighborhoods with aging or aged populations on the outskirts of the urban core. As outlined above, these populations are underserved by both the private and public projects currently spread across the city.

References

Christensen S, Malberg Dyg P, Allenberg K (2019) Urban community gardening, social capital, and "integration" – a mixed method exploration of urban "integration-gardening" in Copenhagen, Denmark. Local Environ 24(3):231–248. https://doi.org/10.1080/13549839.2018.1561655

Dennis M, Cook PA, James P, Wheater CP, Lindley SJ (2020) Relationships between health outcomes in older populations and urban green infrastructure size, quality and proximity. BMC Public Health 20(1):626–626. https://doi.org/10.1186/s12889-020-08762-x

Evans P (2021) We'll all be paying a lot more for food next year, says Canada's food price report. CBC News. Retrieved March 21, 2022, from https://www.cbc.ca/news/business/food-price-report-charlebois-1.6277590

Fuenfschilling L (2017) A review of "retrofitting cities: priorities, governance and experimentation", edited by Mike Hodson and Simon Marvin. Int J Hous Policy 17(1):160–161. https://doi.org/10.1080/19491247.2016.1267458

Fuenfschilling L, Frantzeskaki N, Coenen L (2019) Urban experimentation & sustainability transitions. Eur Plan Stud 27(2):219–228. https://doi.org/10.1080/09654313.2018.1532977

Gale DE (1979) Middle class resettlement in older urban neighborhoods. J Am Plan Assoc 45(3):293–304

Håkansson I (2019) Urban sustainability experiments in their socio-economic milieux: a quantitative approach. J Clean Prod 209(Journal Article):515–527. https://doi.org/10.1016/j.jclepro.2018.10.095

Håkansson I (2021) Goals and persistence of sustainability experiments in divergent urban contexts: urban agriculture and a geodemographic classification in London. Local Environ 26(6):736–753. https://doi.org/10.1080/13549839.2021.1916900

Harper ET (2020) Ecological gentrification in response to apocalyptic narratives of climate change: the production of an immuno-political fantasy. Int J Urban Reg Res 44(1):55–71. https://doi.org/10.1111/1468-2427.12842

Kerstein R (1990) Stage models of gentrification: an examination. Urban Aff Q 25(4):620–639. https://doi.org/10.1177/004208169002500406

Kleinschroth F, Kowarik I (2020) COVID-19 crisis demonstrates the urgent need for urban greenspaces. Front Ecol Environ 18(6):318–319. https://doi.org/10.1002/fee.2230

Kronsell A, Mukhtar-Landgren D, Institutionen för globala studier, Göteborgs universitet, Gothenburg University, Samhällsvetenskapliga fakulteten, School of Global Studies, & Faculty of Social Sciences (2018) Experimental governance: the role of municipalities in urban living labs. Eur Plan Stud 26(5):988–1007. https://doi.org/10.1080/09654313.2018.1435631

Lehrer U, Wieditz T (2009) Condominium development and gentrification: the relationship between policies, building activities and socio-economic development in Toronto. Can J Urban Res 18(1):140–161

Moos M (2016) From gentrification to youthification? The increasing importance of young age in delineating high-density living. Urban Stud (Edinburgh, Scotland) 53(14):2903–2920. https://doi.org/10.1177/0042098015603292

Moos M, Walter-Joseph R, Biglieri S (2017) Still detached and subdivided?: suburban ways of living in 21st-century North America. Jovis Verlag GmbH. https://go.exlibris.link/kvTVmw25

Nicholls E, Ely A, Birkin L, Basu P, Goulson D (2020) The contribution of small-scale food production in urban areas to the sustainable development goals: a review and case study. Sustain Sci 15(6):1585–1599. https://doi.org/10.1007/s11625-020-00792-z

Patton J (2021) Toronto experiences record population loss as more move away from city amid COVID-19 pandemic. Globalnews.ca. Retrieved March 21, 2022, from https://globalnews.ca/news/7575913/toronto-housing-canada-population-coronavirus/

Rice JL, Cohen DA, Long J, Jurjevich JR (2020) Contradictions of the climate-friendly city: new perspectives on eco-gentrification and housing justice. Int J Urban Reg Res 44(1):145–165. https://doi.org/10.1111/1468-2427.12740

Singleton AD, Longley P (2015) The internal structure of Greater London: a comparison of national and regional geodemographic models. Geo Geogr Environ 2(1):69–87. https://doi.org/10.1002/geo2.7

Smith N (2002) New globalism, new urbanism: gentrification as global urban strategy. Antipode 34(3):427–450. https://doi.org/10.1111/1467-8330.00249

Toronto and Region Conservation Authority (TRCA) (2022) Urban agriculture. Retrieved September 2, 2022, from https://trca.ca/conservation/urban-agriculture/

Index

A
Agricultural land, 8, 9, 11, 12, 17–20, 23, 24, 31, 186, 219
Airbnb, 195–213
Apartment rentals, 195

B
Bivariate, 140, 147–149, 151, 152, 157, 158, 166, 168–170, 205, 208, 209, 212

C
Census, 18, 42–45, 47, 48, 52–56, 58, 68, 71, 93, 98–102, 105, 109, 116, 118–120, 122, 130–133, 140, 141, 146–173, 199, 200, 202–204, 206–208, 222, 224, 225
COVID-19, 35–44, 46, 47, 49–63, 202

E
Explanatory models, 140, 150–152, 154, 156–158, 165, 166, 168, 169, 171, 172

F
Food retail, 36–38, 44–46, 58, 60, 62

G
Geodemographic, 37, 217–237
Geographically weighted regression (GWR), 37, 42, 49, 50, 57, 59, 60, 62, 98, 115, 116, 118–121, 129–132, 172
Geography, 9, 23, 37, 76, 110, 152
GIS, 46, 110, 111, 151

H
Hierarchal cluster, 140, 148, 149, 152–154
Historical overview, 7

I
Inadequate dwellings, 139–173
Informal caregiving, 97–132

L
Landsat imagery, 25
Landscape metrics, 9, 17, 26, 29, 30, 32
Land-use, 1–14, 17, 22–24, 26, 29–32, 71, 74, 89–91, 93, 184–186, 189
Land use change, 4, 7–10, 12, 14, 17, 25, 32
Land-use/land cover change, 1, 4, 9, 12

N
Niagara Region, 17–32

O
Ontario, 18–21, 31, 42, 51, 59, 66, 95, 97–99, 101, 102, 108–112, 116, 118, 120, 121, 123, 125, 127, 129, 130, 132, 181–192, 195–197, 202, 217, 219, 220, 235

E. Vaz, *Regional and Urban Change and Geographical Information Systems and Science*, Advances in Geographic Information Science,
https://doi.org/10.1007/978-3-031-24731-6

P
Policy analysis, 181–192

R
Regional planning, 6, 12, 31, 32
Rental markets, 195–202, 211–213
Repairs, 139, 142, 144, 145, 147, 148, 151, 154, 157, 158, 161–163, 165, 166, 168, 169, 171, 172, 184, 187

S
Senior, 38, 97–102, 105–111, 116, 118–121, 123, 125, 130–133, 143
Short term rentals, 195–197, 199–202, 210–213
Spatial analysis, 5, 9, 17, 36, 46, 58, 60, 65–94, 109, 111, 115, 132, 139, 149, 152, 202, 203
Spatial and temporal analysis, 195
Spatial autocorrelation, 41, 46, 48, 50, 52, 66–69, 75, 76, 78–83, 93, 112, 113, 118, 152, 158–167, 202, 204, 205, 208, 209
Spatial decision support, 37
Stepwise regression, 76, 77, 89–93, 140, 148–157, 168, 171
Stepwise regression modeling (SRM), 89
Strategy, 1, 3, 4, 10, 13, 36, 39, 40, 58, 60, 63, 94, 139, 145, 146, 152, 173, 184, 188–192, 222
Sustainability, 4, 6, 9–12, 23, 183, 221–224

T
Toronto, 20, 31, 32, 35–63, 66, 69–76, 78–83, 85–89, 91, 93–95, 111, 139, 141, 146–148, 152, 156, 157, 159–173, 195–197, 199–212, 217–237

U
Urban agriculture, 217–237
Urbanization, 1, 3, 5, 9–12, 30, 93, 217
Urban sprawl, 4, 9, 10, 12, 14, 17, 18, 23, 24, 30, 31

V
Vehicle collisions, 65, 66, 70, 74, 78
Volunteered geographic information (VGI), 12, 13

W
Wetlands, 181–192